Two Phase Separation in the T-Junction

Faheem Ejaz · William Pao · Hafiz Muhammad Ali

Two Phase Separation in the T-Junction

Springer

Faheem Ejaz
Department of Mechanical Engineering,
Faculty of Engineering
King Fahd University of Petroleum
and Minerals
Dhahran, Saudi Arabia

Hafiz Muhammad Ali
Department of Mechanical Engineering
and Interdisciplinary Research Center
for Sustainable Energy Systems (IRC-SES)
King Fahd University of Petroleum
and Minerals
Dhahran, Saudi Arabia

William Pao
Department of Mechanical Engineering
Universiti Teknologi PETRONAS
Seri Iskandar, Perak, Malaysia

ISBN 978-981-99-9353-6 ISBN 978-981-99-9354-3 (eBook)
https://doi.org/10.1007/978-981-99-9354-3

This Springer imprint is published by the registered company Springer Nature Singapore Pte Ltd.
The registered company address is: 152 Beach Road, #21-01/04 Gateway East, Singapore 189721, Singapore

Paper in this product is recyclable.

Acknowledgements

Hafiz Muhammad Ali acknowledges the support of King Fahd University of Petroleum and Minerals, Dhahran 31261, Saudi Arabia.

Contents

Abbreviations

3D	Three Dimensional
CFD	Computational Fluid Dynamics
CSF	Continuum Surface Force
DR	Diameter Ratio
FVM	Finite Volume Method
GCI	Grid Convergence Index
LCT	Liquid Carryover Threshold
MSR	Mass Split Ratio
PLC	Peak Liquid Carryover
PS	Phase Separation
SE	Separation Efficiency
SEE	Standard Error of Estimate
VOF	Volume Of Fluid
VR	Velocity Ratio

Nomenclature

D	Diameter of the pipe (m)
DR_1	Upstream diameter ratio
DR_2	Downstream diameter ratio
V_a	Air superficial velocity (m/s)
V_G	Gas superficial velocity (m/s)
V_w	Water superficial velocity (m/s)
V_L	Liquid superficial velocity (m/s)
F_a	Volume fraction of air in the side arm
F_w	Volume fraction of water in the side arm
F_G	Volume fraction of gas in the side arm
F_L	Volume fraction of liquid in the side arm
W	Volumetric flow rate (m^3/s)
W_G	Volumetric flow rate of gas (m^3/s)
W_L	Volumetric flow rate of liquid (m^3/s)
$T_{1\text{-}5}$	Suddenly expanded T-junction design (1–5)
S	Dimensionless parameter
ρ	Density (kg/m^3)
ν	Kinematic viscosity (m^2/s)

Subscripts

1	Main arm
2	Run arm
3	Side arm
a	Air
w	Water
G	Gaseous phase
L	Liquid phase

List of Figures

List of Tables

Chapter 1
Introduction

Natural gas is prominently growing as the most vital way to produce excessive amount of power production. There are multiple reasons for such a verdict. Firstly, other sources of energy are less clean than natural gas itself and secondly, there is industrial significance in using natural gas as the energy source in terms of economy and environment. It was mentioned by [1] that the consumption of natural gas as an energy source almost doubled in the years of 1999 and 2020, both domestically and industrially.

Transportation of natural gas through piping networks of the most favourable option because of very low maintenance costs. This is true when the place where the gas is to be transported is not far. One of the biggest piping network layouts is owned by the USA with almost two hundred thousand miles of transmission lines as well as twelve hundred stations to accommodate necessary compressions [2]. Figure 1.1 represents the main piping networks in the US for gas distribution.

The key parameters of such complex flow systems are individual phase superficial velocities, liquid holdup, and pressure drop. The calculation of these parameters involves representing the networks in the form of graphs.

When it comes to the two-phase flow, gas phase petroleum mixtures are extracted from underground with the help of special pipe extraction facilities. Two-phase can be the cause of two main reasons. Whether it is due to the gas and liquid flowing side by side during petroleum extraction or due to the own selection of field operator or unforeseen condensation of denser petroleum fluids in the pipelines. The two-phase flow consists of two distinct phases traveling side by side. The most common type of flow is liquid–gas flow. Such flows can be exemplified as water–oil, oil–gas, or water–gas flows. Sometimes, despite the fact that the inlet is only a single-phase stream, the denser hydrocarbons liquefy due to condensation, or the water is found in emulsion along with other petroleum products. The extent of condensation depends upon the density of the working fluids, bore condition, and type of the source of crude oil. When the mixture of phases is transported to the onshore sites, it is often desired to separate them into distinct phase streams to make use of them. From the

F. Ejaz et al., *Two Phase Separation in the T-Junction*,
https://doi.org/10.1007/978-981-99-9354-3_1

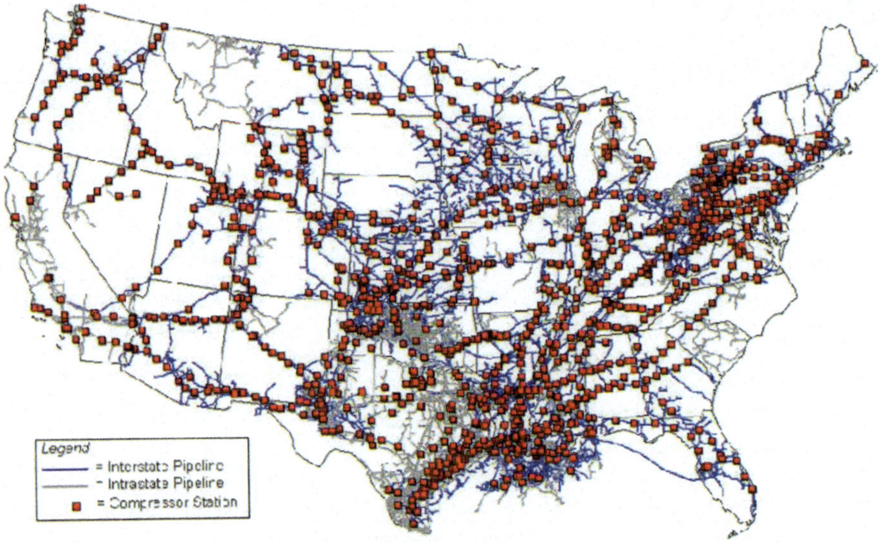

Fig. 1.1 Pipeline and compression station systems in the USA

extraction point, it is favorable to keep both phases flowing together due to the lesser availability of space.

There are two major types of T-junctions. Branching T-junction (Fig. 1.2) plays a vital role in partial phase separation, and it is an essential part of many bigger and smaller piping networks in the petroleum industries. Flow through a T-junction experiences pressure drops especially at the branching arm as well as phase maldistribution. This consequently results in poor outflow stream quality as well as hazards to downstream equipment such as gas turbines. In addition, to design a good performance piping network it is a serious concern to address the phase separation behavior in the T-junction and quantitatively analyze it in terms of key performance indices such as separation efficiencies, peak liquid carryovers, or liquid carryover threshold.

Some of the practical examples where a two-phase flow is identified are as follows:

- Power plants where streams of liquid and gas are to be transported to different locations [3].
- Water reactors in the nuclear power plants for cooling purposes [4] and related break discharges [5]. When the pipelines burst abruptly the water evaporates due to higher pressure reductions and causes a two-phase flow in cooling networks. This is also known as Loss of Coolant Accidents (LOCA) [6].
- Condensation formulation in the complex piping systems [7].
- Enhanced oil recovery operations (EOR) [8, 9].
- Phase separators like gas scrubbers [10, 11].
- Two-phase refrigerant flows in multiple heating ventilation and air conditioning systems [12].

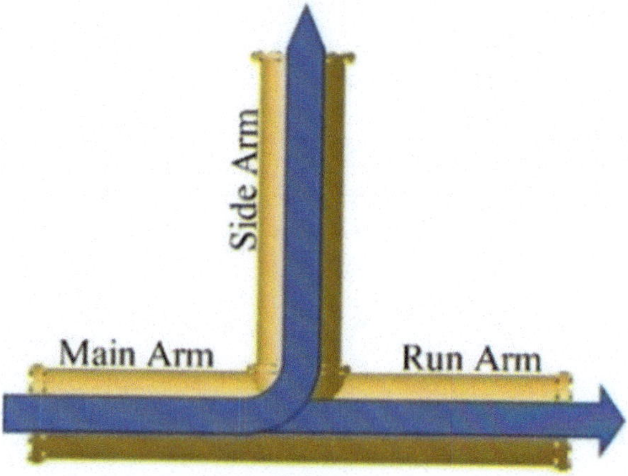

Fig. 1.2 Branching T-junction

It is imperative to address that in the literature, the term "phase separation" is consistently used to quantify the individual phase redistribution behavior in the T-junction. On the other hand, "phase distribution" indicates the capability of one phase to cover the specific areas of the pipe cross sections (e.g., water covers bottom part due to the gravity and higher density). The definition of "phase split" is the formation of the second working fluid phase because of pressure drop or temperature changes.

1.1 Single Phase Flow Systems

Flow systems or networks are known as organized systems of nodes and elements of the nodes altogether. The node is the point where two or more pipelines join together to form a junction. The junction term is often used in partial phase separation in the T-junction as well as in complex piping networks. Generally, a complicated piping network involves branching, impacting, parallel, and series combination of pipelines, including other downstream equipment like bends, pumps, and other separation conduits. There are basically two types of networks as shown in Fig. 1.3.

One of the ways to depict the single-phase flow in piping networks is through metrices. Since the matrices can be easily manipulated statistically, the network topology can be indicated for any flow network.

In the rectangular matrix, a number of rows are equal to the number of nodes as well as the number of columns as the total number of pipelines. The formula of elements of the matrix is represented as

$$a_{i,j} = \begin{cases} +1 \text{ if pipe j enters node i} \\ -1 \text{ if pipe j leaves node i} \\ 0 \text{ if pipe j has no connections with node i} \end{cases} \tag{1.1}$$

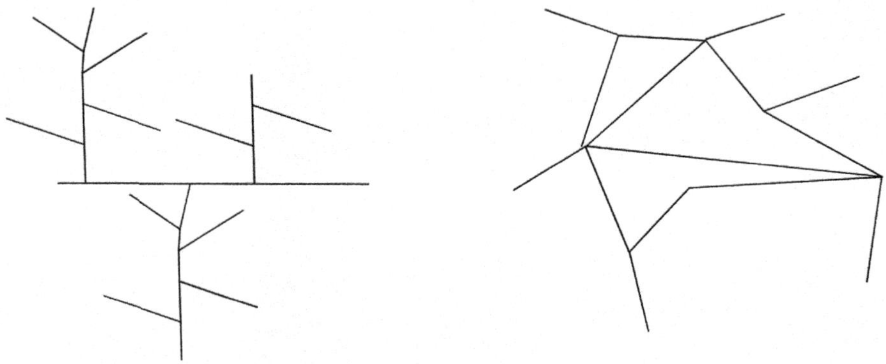

Fig. 1.3 Piping network types (Looped and Loopless)

If the network is looped, the matrix is known as branch loop incidence, and in this matrix the rows are equal to the number of loops and columns as the number of pipelines.

$$b_{i,j} = \begin{cases} +1 \text{ if pipe j has same direction as loop i} \\ -1 \text{ if pipe j has different direction as loop i} \\ 0 \text{ if pipe j is not in loop i} \end{cases} \tag{1.2}$$

Other than these element matrix formulae, [13] has represented five types of methodology for gas flow networks characterization.

1.2 Single Phase Flow in Pipelines and Networks

If the flow parameters such as the flow rate can be calculated the single-phase flow in piping networks of natural gas can be analyzed. In general, pressure drop and flow rate of the natural gas are used to correlate and quantitatively analyze the networks. Single phase flow in networks is solved keeping in view of the following criteria.

- The flow at the inlet of the complex or simple junction should always be equal to the outflow (Kirchhoff's first law).
- Selection of the path by the phase does not affect the energy consumed (Kirchhoff's second law). To be precise, if the flow path lines are in a loop, the fluid particle starting and ending at the same point will have the same energy as it was in the initial stage.

For the actual calculations of single-phase flow parameters, phase in extraction bores and pipes are assumed to flow in one direction especially when the change in parameters in the axial direction is to be calculated. For the flow in three dimensions, the flow parameter values are volume averaged over the pipe cross-sectional area. This subsequently reduces the Navier-Stokes equations to one dimension only. Nevertheless, viscous forces are only dominant near the pipe wall known as the boundary layer. Since the boundary layer thickness is smaller than the total pipe length, the volumetric forces can be neglected when compared to the general forces responsible for the fluid flow. According to the application, the extra terms of heat transfer, viscosity, or shear forces can be introduced.

1.3 Two-Phase Flow in Pipelines

Two-phase flow is not only prominent in many offshore industrial applications such as petroleum, chemical, nuclear, and heat transfer applications but also in complex piping networks for domestic transportation of a phase. This turns out to be a bigger challenge for the industries to solve the unpredictable behavior of phase redistribution. Usually, when two or more phases are being transported simultaneously, an abundance of turbulent phenomena is experienced which makes it very difficult to analyze the phase separation through experimentation or computation. When two phases flow, due to the fact that they have different densities, inertia, and compressibility, the phases form flow regimes also known as flow patterns depending upon the flow characteristics. Flow patterns and their identification will be discussed in the next chapter in detail.

1.3.1 Flow Characteristics

Flow pattern usually changes by the slightest change of phase superficial velocity, inlet quality, dimensions of the pipes, densities, the orientation of the pipe, and other flow characteristics. Many authors have worked on identifying the flow pattern based on these parameters with different methods. For instance, [14] reviewed and correlated the flow regimes observed in the literature and concluded the flow regime types as shown in Fig. 1.4. Further after identifying the flow patterns, the flow parameters readings were recorded and plotted in a flow pattern map as represented in Fig. 1.5. The flow rates of gas and air were used as x- and y-coordinates, respectively.

Two-phase flow pattern has been an important area to explore for the past 30 years. Usually, the models are divided into two main categories.

- Homogenous equilibrium model

  This model considers that when two phases are flowing side by side, the flow properties are averaged for both phases.
- Pseudo-homogenous model

  One example includes the Drift-Flux Model for which slip condition is introduced which ultimately relies on the flow characteristics.

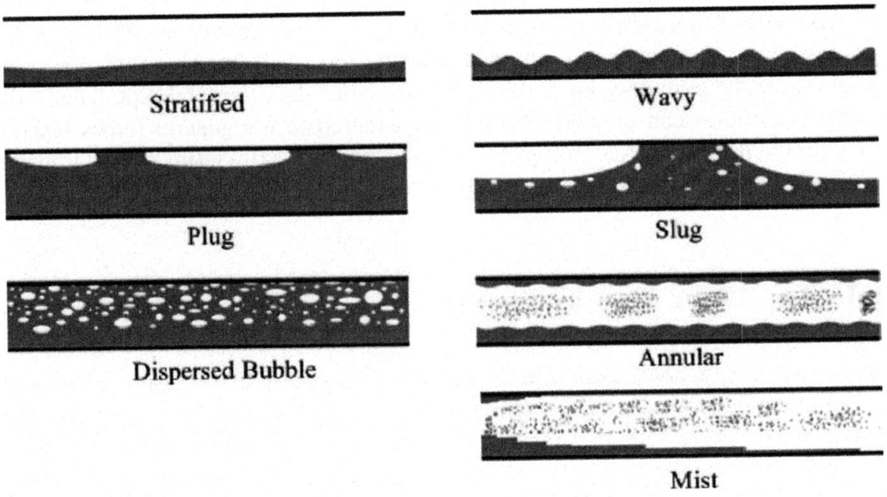

Fig. 1.4 Types of flow regimes [14]

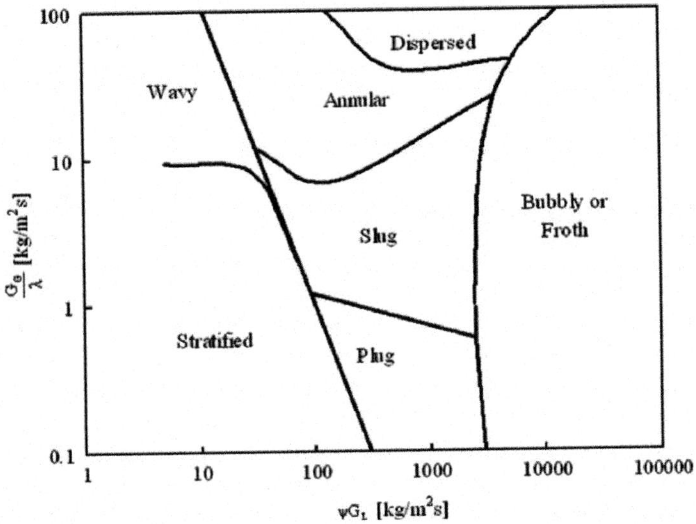

Fig. 1.5 Flow pattern map [14]

- Heterogeneous models

  This includes two types of models such as separated model and two-fluid model. In a separated model, different zones are devised for each flow transport while in two-fluid model in each flow zone, two phases can flow simultaneously.

It is a fluid dynamic phenomenon in which pressure drop can occur at different points of the junction during transport. Therefore, it is imperative to estimate the pressure loss due to different factors such as friction or elevation. [15, 16] were among the first few researchers who predicted the pressure loss in horizontally inclined pipes in the years 1948 and 1949.

One of the most famous empirical prediction models till date was presented by [17]. Further, [18] gathered data from oil extraction facilities in Saudi Arabia. In their model, if the working fluids such as oil and gas are known along with their heat capacities, densities, flow rates, etc., the pressure drop, temperature gradient, volume fractions, flow pattern, and gas velocity required to extract liquid completely from the riser can be calculated.

1.3.2 Two-Phase Flow Variables

To study the two-phase flow, the variables, such as pressure, temperature, liquid holdup, volume fraction, mixture velocity, etc., need to be known through governing equations.

- Liquid holdup

  It is defined as the volume fraction of liquid in a specific volume.

$$F_L = \frac{V_L}{V_L + V_G} \tag{1.3}$$

  whereas F_L, V_L, and V_G represent fraction of liquid, volume of liquid, and volume of gas phase, respectively.

- Void fraction

  It is also known as the volume fraction of gas in any volume.

$$F_G = \frac{V_G}{V_L + V_G} \tag{1.4}$$

  In two-phase flow where gas and liquid flow simultaneously, the total volume fraction will always be equal to unity.

$$F_L + F_G = 1 \tag{1.5}$$

- Phase superficial velocity:

  It can be calculated as the ratio of the volume flow rate to the area of the cross section.

$$v_k = \frac{W_k}{A} \tag{1.6}$$

whereas v_k, W_k, and A represent the velocity of the kth phase, flow rate of the kth phase, and area of the cross section, respectively.

- Velocity of the mixture

$$v_{mix} = \frac{W_G + W_L}{A} \tag{1.7}$$

- Slip velocity

This term is also known as the relative velocity of both phases.

$$v_{slip} = v_G - v_L \tag{1.8}$$

- Drift velocity

It is defined as the velocity of one phase as compared to a mixture.

$$(v_k)_{drift} = v_k - v_{mix} \tag{1.9}$$

- Drift flux

It is defined as the area flow rate of a phase relative to mixture's superficial velocity.

$$J_k = F_k(v_k - v_{mix}) \tag{1.10}$$

- Quality of the flow

Quality of gas in any fluid stream is as follows.

$$x = \frac{\dot{m}_G}{\dot{m}_G + \dot{m}_L} \tag{1.11}$$

whereas $\dot{m}$ indicates mass flow rate.

1.4 Phase Separation Background

Two-phase flow is often experienced in many petroleum, chemical, and nuclear reactor applications. In most of the cases, the separation of different phases is required to handle each phase differently and make use of them. There are minor applications, where phase separation is only required for easier handling of a specific phase such as gas transportation. For this purpose, many phase separation conduits are used. Bigger and bulkier phase separators are utilized that use centrifugal phenomena or the effect of gravity to separate different density phases. In the production stage, the separators are considered as a bottleneck because of the following reasons:

- It is required to keep the mixture of phases in the separators for an extensive period for phase separation.
- The separators demand larger areas for the installations.
- The material used in the phase separators is flammable which is even riskier.

Therefore, to address these issues industrialists and researchers agreed upon using a simpler yet effective phase separation conduit known as T-junction. T-junction is utilized only for partial phase separation, and it is installed just upstream of the other separators such as gas scrubbers. Since T-junction initially separates the phases up to a larger extent, it further reduces the workload on downstream equipment used for separation. Two of the biggest advantages of T-junction are its simple design (Two pipes intersect with each other to form a junction) and very low maintenance expenses. Conte [19] in his research discussed that Leonardo da Vinci was the first researcher to observe phase redistribution in the complex piping T-junctions in sixteenth century. Later in twentieth century, Acrivos [20] studied the phase separation in complex pipe systems analytically.

Even though T-junction is very simple in design and requires very little maintenance costs, it is very unpredictable how it will behave for a particular inlet parametric condition for a specific working fluid. One of the reasons is that the phase separation in the T-junction depends upon eight distinct geometric factors as well as other flow properties. The geometric parameters include diameters of the main, run arm and side arm, length of the side arm, diameter ratio of the side arm, angle of inclination of the main and side arm, and radius of curvature of the junction. Flow parameters include densities of the working fluids, type of flow regime, inlet gas and liquid superficial velocities, and outflow flowrate weightage. Figure 1.6 illustrates the factors effecting the phase separation.

1.5 Phase Maldistribution in Two-Phase Flow

In general, in two-phase flow, two working fluids having different phases have different densities. It is evident that a phase with higher mass will have higher density and is more prone to gravity. Therefore, it stays at the lower part of the pipes and flows along the characteristic length. The phase with higher density will also have higher inertia so it will require some force to push the liquid into the branch arm for phase maldistribution to occur. Similarly for phases with lower mass or for the gas phase, the densities are quite lower, leading to the gas phase covering the top parts of the pipes and ultimately separating through the branch arm. This separated gas can further be used for power production through gas turbines. In practice, the phases unequally distribute among the T-junction outlets, known as phase maldistribution [21–23]. In the case of phase maldistribution, the liquid takes off from the side arm due to a higher pressure drop in the branch created by the gas phase. Downstream equipment such as gas scrubbers and gas turbines are not designed or manufactured to accommodate higher proportions of the liquid, therefore, the whole separation

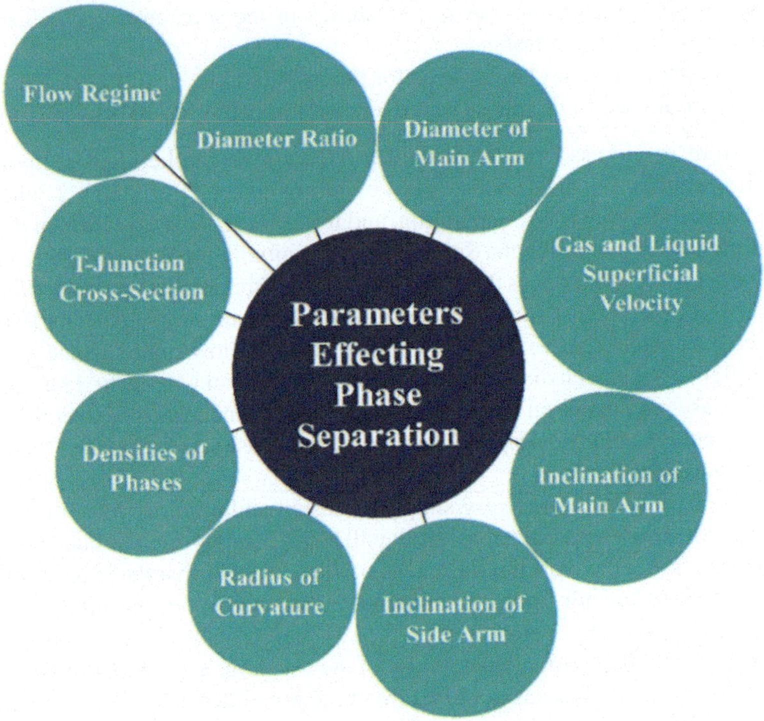

Fig. 1.6 Variables effecting the separation performance of the T-junction

plant experiences frequent shutdowns or causes wear and tear of the downstream equipment.

Oranje [7] reported in his research that while he was working on piping networks in the Netherlands, he discovered that one outlet of the pipes had dry gas and other outlets had gas condensates in them. This invoked the concept that there must be an unequal phase redistribution at the junction causing this behavior. His findings are mentioned multiple times in the literature [19, 24] because of his introduction to the fundamentals of phase separation.

The phase separation behavior is still a matter of concern since the very first observation by Leonardo da Vinci. He investigated and reported the phase vortex and recirculating phase at the junction. In general, it is very difficult to predict the phase separation performance. The problem and uncertainty complicate where there are sharp bends and changes in flow directions like in the case of T-junction. In the sharp typical 90-degree bend in the T-junction, the pressure abruptly changes among the outlet arms of the network. This poses a sudden increase and drop in the pressure which causes unforeseen liquid take offs.

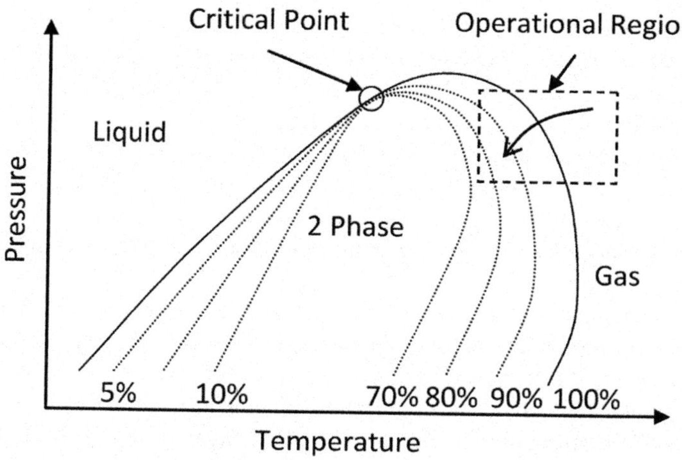

Fig. 1.7 Phase envelope graph of natural gas

1.5.1 Route Selectivity for Two-Phase Flow

In the transportation of natural gas for domestic or industrial use, if there are higher density wet gas stream, the phase envelope of the mixture can provide a restriction to the pipe's operational region as shown in Fig. 1.7. If the region of operation is inside the phase envelope for natural gas, gas will condense and work as a secondary phase known as gas condensate. If the operational region is partially coinciding with the phase envelope, the gas will only condense at the cross sections of the pipes for which pressure drops.

Even if the inlet conditions are entirely single phase, it is highly possible that if the pressure in the pipe drops below the dew point of the phase, it will condense the gas and convert the single-phase flow into the two-phase flow.

The presence of condensate in single-phase flow acts as a secondary phase and provides varying hydraulic pressure changes along the pipe length. This would seriously affect the transportation of primary working fluid. Condensates also change the overall mass transfer and may vary the composition of the mixture because of fluctuations in flow characteristic parameters such as flow rate, density, and volume. Consequently, the outlet product quality or the transportation of a primary phase is significantly affected.

If the two-phase flow is because of the condensate presence, following is the ways to quantitatively analyze.

- Empirical or theoretical prediction of two-phase flow redistribution along the pipe.
- Pressure-drop, temperature gradient, and separation performance calculation in the T-junction.

It is evident to say that the flow distribution in the T-junction can be treated as the electric circuit resistances. The resistance to flow can be in the form of geometrical factors (diameters, angle of inclination, length, roughness). For instance, in the gas of gradual divergence in the T-junction, phase separation is only affected by the pressure drop and restriction to the flow due to divergence, independent of other geometrical changes. Just like the electrical circuits, if the flow resistance on the run arm and branch arm is equal, the phases will split evenly. The reason for this behavior is that if the resistance on both outlets is the same, the pressure drop will also be the same. Therefore, phases will be evenly split. This held true for gas condensate and natural gas flow simultaneously when the difference in density and gravity effect is not very prominent.

On the other hand, in two-phase flow such as air–water flow, the difference in densities is very high such that the phase separation becomes unpredictable at the junction. Because of differences in densities, wall frictions for a specific phase, and drag, both phases can behave differently for the same gradients of pressure [25–27].

In an ideal situation, the gas being the lighter phase takes off through the side arm if we consider the case of an upward 90-degree side arm branching T-junction. The liquid due to higher density and momentum stays in the lower part of the pipe and flows directly through the run arm. The higher density phase is very sensitive to pressure changes such that if the lighter phase flow rate in the branch arm is increased, it significantly drops the pressure. This pressure drop further dominates the inertia and gravity effect of the heavier phase (liquid). Therefore, a high volume fraction of the liquid diverts into the side arm causing issues for the downstream equipment and reducing the branching outlet quality. This phenomenon is called phase maldistribution or the sudden change in the phase separation behavior was also reported as the flip-flop effect as shown in Fig. 1.8.

To efficiently predict the phase separation, it is imperative to analyze all parameters associated with the phase separation influence. These parameters may include

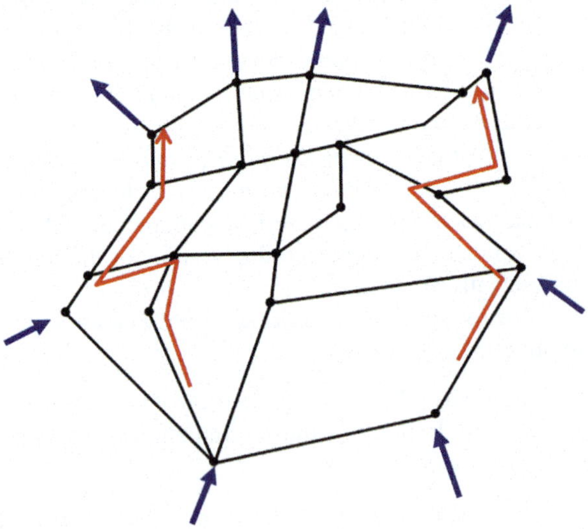

Fig. 1.8 Rout preference in the piping network

densities, flow rate of each phase, diameter of arms, angle of inclinations, outlet back pressure, and other influencing factors. The slightest change in any of the parameters can entirely change the phase separation performance of the T-junction.

References

1. M. Balat, World natural gas (NG) reserves NG production and consumption trends and future appearance. Energy Sources. **27**, 921–929 (2005). https://doi.org/10.1080/00908310490449027
2. U.S. Energy Information Administration, Natural gas compressor stations on the interstate pipeline network: developments since 1996, Energy Inf. Adm. Off. Oil Gas. 1–12 (2007). http://www.eia.gov/pub/oil_gas/natural_gas/analysis_publications/ngcompressor/ngcompressor.pdf
3. A. Fauda, E. Rhodes, Two-phase annular flow stream division in a simple tee. Trans. Instn. Chem. Eng. **52**, 354–360 (1974)
4. R.T. Lahey, Current understanding of phase separation mechanisms in branching conduits. Nucl. Eng. Des. **95**, 145–161 (1986). https://doi.org/10.1016/0029-5493(86)90043-9
5. C. Smoglie, J. Reimann, U. Muller, Two phase flow through small breaks in a horizontal pipe with stratified flow. **99**, 117–130 (1987)
6. N.E. Todreas, M.S. Kazimi, *Nuclear Systems, vol I, Thermal Hydraulic Fundamentals*, 3rd edn. (2021)
7. L. Oranje, Condensate behavior in gas pipelines is predictable. Oil Gas J. **71**, 39–44 (1973)
8. E.L. Berger, K.W. Kolthoff, J.L.G. Schrodt, S.L. Long, J.C. Pauley, The SpliTigatorTM: a device for the mitigation of phase splitting. Int. Therm. Oper. Heavy Oil Symp. SPE-37516-MS (1997). https://doi.org/10.2118/37516-MS.
9. J. Jones, R.L. Williams, A two-phase flow-splitting device that works. SPE Prod. Facil. **8**, 197–202 (1993). https://doi.org/10.2118/21532-PA
10. D.P. Margaris, T-junction separation modelling in gas-liquid two-phase flow. Chem. Eng. Process. Process Intensif. **46**, 150–158 (2007). https://doi.org/10.1016/j.cep.2006.05.007
11. S. Rea, B.J. Azzopardi, The split of horizontal stratified flow at a large diameter T-junction. Chem. Eng. Res. Des. **79**, 470–476 (2001). https://doi.org/10.1205/026387601750282409
12. S.J. Tae, K. Cho, Two-phase flow distribution and phase separation through both horizontal and vertical branches. KSME Int. J. **17**, 1211–1218 (2003). https://doi.org/10.1007/BF03016516
13. A. Osiadacz, *Simulation and Analysis of Gas Networks*. (Gulf Publishing Company,Houston, TX, United States, 1987). https://www.osti.gov/biblio/5141539
14. L. Cheng, G. Ribatski, J.R. Thome, Two-phase flow patterns and flow-pattern maps: fundamentals and applications. Appl. Mech. Rev. **61** (2008). https://doi.org/10.1115/1.2955990
15. R.C. Martinelli, D.B. Nelson, Prediction of pressure drop during forced-circulation of boiling water. Trans.ASME. **695** (1948)
16. R.W. Lockhart, R.C. Martinelli, Proposed correlation of data for isothermal two-phase, two-component flow in pipes. Chem. Eng. Prog. 38–48 (1949)
17. D.H. Beggs, J.P. Brill, A study of two-phase flow in inclined pipes. J. Pet. Technol. **25**, 607–617 (1973). https://doi.org/10.2118/4007-PA
18. J.P. Brill, T.R. Sifferman, B. Samaroo, S. Arirachakaran, Simulation of a major oilfield gas-gathering pipeline system. J. Energy Resour. Technol. **107**, 397–401 (1985). https://doi.org/10.1115/1.3231208
19. G. Conte, B.J. Azzopardi, Film thickness variation about a T-junction. Int. J. Multiph. Flow **29**, 305–328 (2003). https://doi.org/10.1016/S0301-9322(02)00130-1
20. A. Acrivos, B.D. Babcock, R.L. Pigford, Flow distributions in manifolds. Chem. Eng. Sci. **10**, 112–124 (1959). https://doi.org/10.1016/0009-2509(59)80030-0

21. B.J. Azzopardi, S. Rea, Phase separation using a simple T-junction, in *Proceeding of the SPE Annual Technical Conference and Exhibition?* (2000), pp 355–362. https://doi.org/10.2523/63040-ms
22. W. Pao, F.M. Hashim, L.H. Ming, Computational analyses of passive wet gas separation in branched piping. MATEC Web Conf. **13**, 1–5 (2014). https://doi.org/10.1051/matecconf/20141303009
23. S.F. Chien, Phase splitting of wet steam in annular flow through a horizontal branching tee. SPE Prod. Facil. **11**, 83–88 (1996). https://doi.org/10.2118/28542-PA
24. K.C. Hong, two-phase flow splitting at a pipe tee. JPT J. Pet. Technol. **30**, 290–296 (1978). https://doi.org/10.2118/6530-PA.
25. J.D. Ballyk, M. Shoukri, A.M.C. Chan, Steam-water annular flow in a horizontal dividing T-junnction. Int. J. Multiph. Flow **14**, 265–285 (1988). https://doi.org/10.1016/0301-9322(88)90044-4
26. R.T. Hwang, S.T, Soliman, H.M, Lahey Jr, R. T. LAHEY Jr, Int. J. Mul. **14**, 439–458 (1988)
27. O. Shoham, S. Arirachakaran, J.P. Brill, Two-phase flow splitting in a horizontal reduced pipe tee. Chem. Eng. Sci. **44**, 2388–2391 (1989). https://doi.org/10.1016/0009-2509(89)85174-7

Chapter 2
Two-Phase Flow in Pipelines

This chapter introduces the concept of multiphase flow as the general term and describes how this came into being and in discussion. We would discuss different flow regimes encountered when gas and liquid flow simultaneously. Specifically, the focus was given to the regimes in horizontal inlets, as these were determined to be important in the industrial setups. Phase separation is a common industrial process. Over the years, this technique has received a great deal of attention with the goal of equipment failure reductions while keeping in view the demand for high efficiencies.

This is relatively complex to actually study the multiphase flow but the compulsion of the industrial application of multiphase flow has motivated many researchers in the area to come up with better and more accurate flow behavior solutions. In the past couple of decades, an abundance of correlations and empirical equations were developed for the ease to reduce experimentation and computational requirements.

2.1 Flow Regimes

When a conduit carries a mixture of both phases, the simultaneous exhibits unique characteristics. While few patterns can be readily identified and recognized, others can be puzzling and resist straightforward categorization. Due to the prevalent presence of two-phase two-phase flow in industrial settings, various experimental setups have been identified to categorize two-phase flows with similar characteristics. They are related to phase distribution and pressure drop, as well as mechanisms that dominate heat and mass transfer. Since the characteristics of hydrodynamic and two-phase flows can be significantly different, it is important to understand and determine the flow regime encountered.

The precise nature of two-phase flow is related to the ratio of phase and the velocity of individual phases. The direction of the pipe also considerably affects the regime because of inertia and gravitational effects. Clearly, many components affect

F. Ejaz et al., *Two Phase Separation in the T-Junction*,
https://doi.org/10.1007/978-981-99-9354-3_2

the two-phase flow system; however, depending on the direction of the pipe, they can be segregated into two separate groups.

2.2 Upward Flow in the Pipelines

The flow patterns for the simultaneous upright surface flow regimes are depicted in Fig. 2.1. The vertical gravity force works upward on the shaft, and the four main flow patterns are stated below.

Bubble Flow: Gas is evenly separated in bubbles of arbitrary size in a uniform length of liquid. On the other hand, the gas bubbles move in an arbitrary way and are observed to merge. It is evident that gas being the lighter phase moves upward as spherical bubbles the size of the tube diameter. Taylor bubbles are often called "Taylor bubbles" and a thin film of liquid can be seen flowing down the wall of the tube along with the rise of the bubbles. The liquid consists of a dispersion of very small bubbles. This flow regime is at times called plug flow.

A flow diagram for a vertical updraft is shown in Fig. 2.1. The axial effect of gravity on vertical updrafts and the four main flow regimes are described below.

Churn Flow: The gas bubbles dissipate in bed flow as it becomes more unique in nature and unpredictable. The flow regimes at high gas velocities are more violent, effervescent, and unique than bubbly flows.

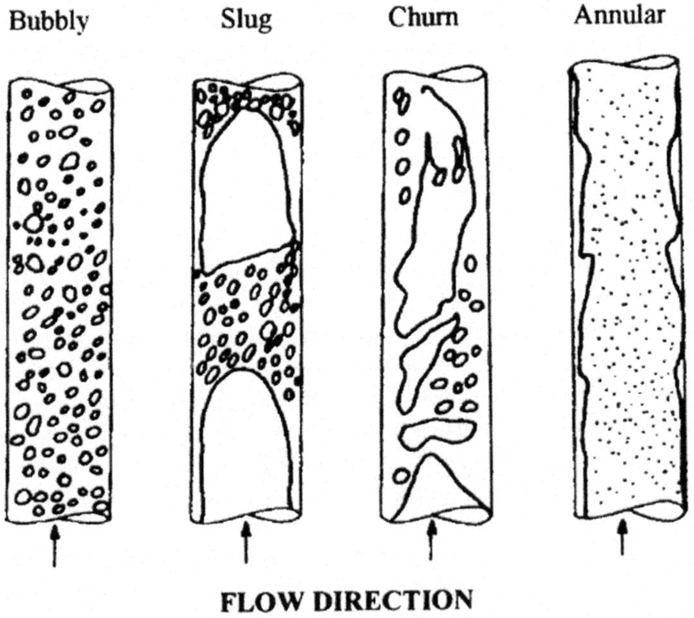

Fig. 2.1 Flow patterns in the vertical pipes of uniform diameter [1]

Annular Flow: Some of the liquid flows as a very fine film of liquid near the walls of the channel and the gas phase flows right at the center of the tube. Some of the liquid discharges into the gas core in the form of fast droplets.

2.2.1 Horizontal Pipe Flows

In this type of flow, gravitation acts at the right angle to the direction of flow. This grounds the two phases to spread in different ways, commonly the dense liquid film streamlined at the bottom half of the pipe. A few of the flow regimes are illustrated in Fig. 2.2.

Bubbly Flow: In this flow regime, air bubbles are distributed inside the liquid film. Nonetheless, gravitation tends to enforce the tiny droplets to collect to the top-half of the pipe, but it is not possible at high superficial gas and liquid velocities, especially if the flow is turbulent enough to segregate the bubbles throughout the pipe core.

Intermittent Flow: A constant layer of liquid in the bottom half of the tube exists and the gas travels on top of it. Waves are generated right along the interface at significantly high gas superficial velocity.

Slug Flow: For upward flow, the flow pattern of wavy. The gas phase moves in the form of large sacks in the upper part of the tube and is isolated by a liquid plug.

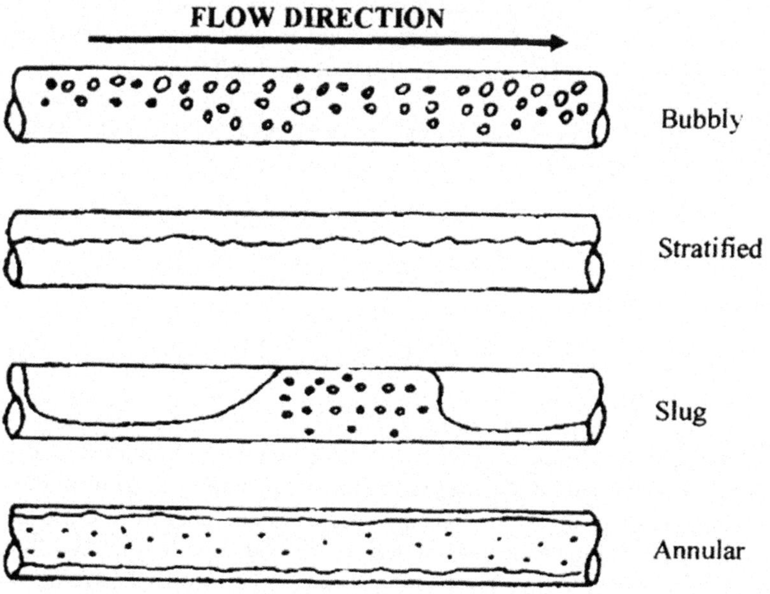

Fig. 2.2 Flow patterns in the horizontal pipes of uniform diameter [1]

Annular Flow: Gas and liquid phase flow in a distinct path as seen in the horizontal flow regime. The gas stays in the core of the pipe whereas the liquid phase sticks to the walls of the conduit.

2.3 Prediction of the Flow Pattern in Horizontal Orientation

Designing a pipeline or equipment requires knowledge of how two-phase flows interact. Bubbly flow has a large area fraction in terms of mass flow rate, making it favorable in many scenarios. Identifying patterns through visual observation is subjective, and high-speed photography is limited in its application. Various instrumental techniques, such as conductance probes and y-ray densitometers, have been developed to measure area fraction or pressure gradient in the pipe. However, the subjective nature of flow pattern determination lack of an extensive data pool to produce a flow pattern map can lead to inaccurate predictions. Additionally, any other geometrical properties or different flow characteristics can all affect flow patterns, indicating the need for more precise predictions. Flow pattern maps, such as those created by Baker [2], Eaton et al. [3], and Taitel and Dukler [4], are the most widely accepted method for predicting two-phase flow regimes, but they have limitations.

In the case of Baker's [2] chart Fig. 2.3, the mass flux of gas and liquid is denoted by G and L, respectively. Two dimensionless parameters, λ and ψ, were introduced by the author as correction factors that can aid in predicting the flow pattern for different working fluids. The equations proposed by the author to govern this prediction are shown in Eqs. 2.1 and 2.2. These correction factors can be used to extend the applicability of the prediction method beyond the specific working fluid used in the original experiment.

$$\lambda = \left[\left(\frac{\rho_G}{\rho_L}\right)\left(\frac{\rho_L}{\rho_W}\right)\right]^{0.5} \tag{2.1}$$

$$\psi = \frac{\sigma_W}{\sigma}\left[\left(\frac{\mu_L}{\mu_W}\right)\left(\frac{\rho_W}{\rho_L}\right)\right]^{\frac{1}{3}} \tag{2.2}$$

The symbols ρ, σ, μ, with subscripts L, G, W, and A denote density, surface tension, and dynamic viscosity, respectively, for liquid, gas, water, and air.

Figure 2.3 displays a distinct line that depicts the transition from one flow pattern to another, although this representation may not accurately reflect real-world scenarios. In actuality, the transition from one flow regime to another is often gradual. To account for this, Mandhane et al. [5] developed a flow regime map by gathering extensive data from the literature. As illustrated in Fig. 2.4, the flow regime transitions are represented by a broader line to accommodate the gradual transition between them.

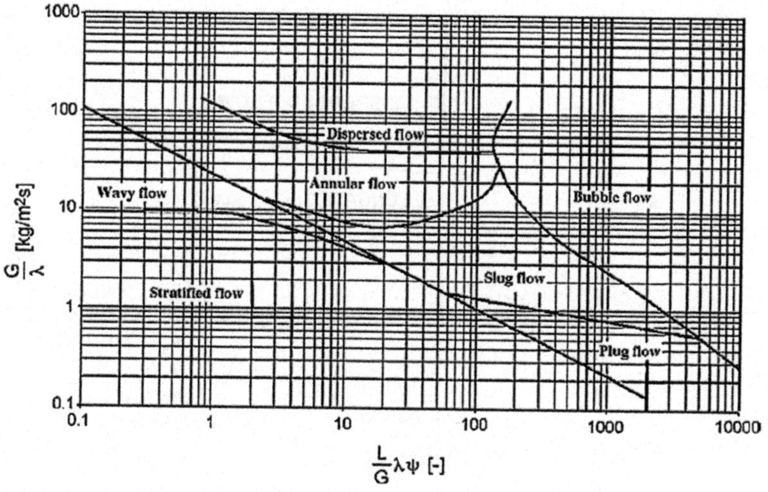

Fig. 2.3 Baker's [2] flow pattern map

Taitel and Dukler's [4] flow regime map, which was produced for specifically 0.127 m of diameter tube, can able to assess the flow pattern for different flow configurations.

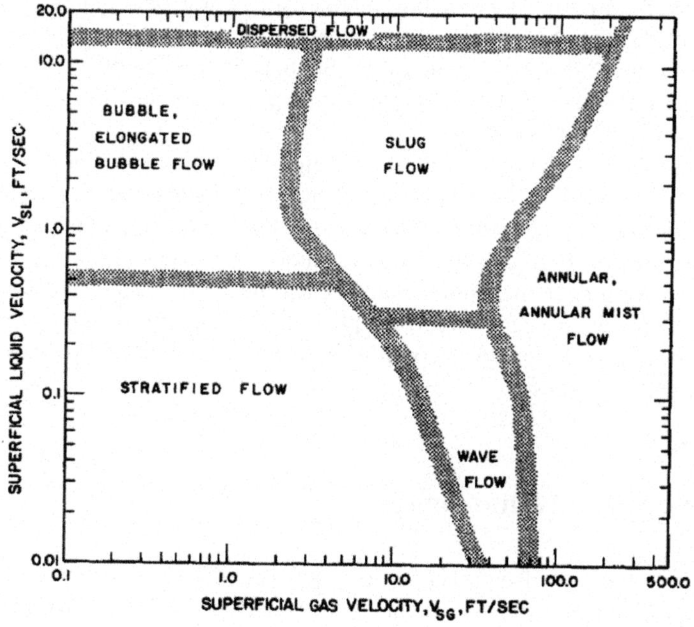

Fig. 2.4 Mandhane et al. [5] flow regime map

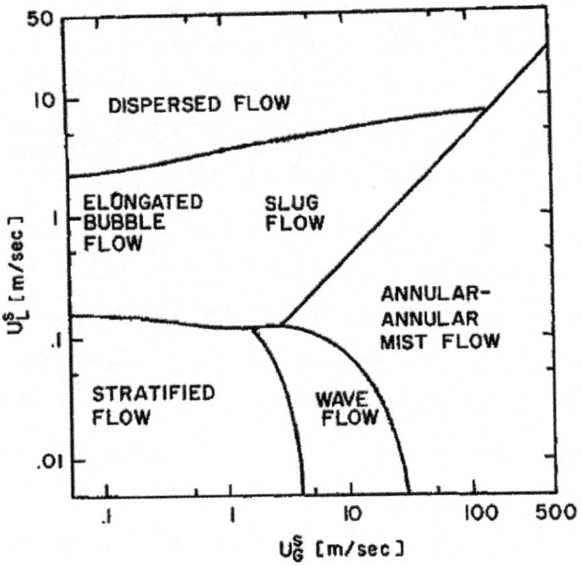

Fig. 2.5 Taitel and Dukler's [4] flow regime map

In recent years, machine learning approaches have also been applied to predict flow pattern in two-phase flow. For instance, Zhang et al. [6] developed a machine learning-based flow pattern prediction model using a convolutional neural network (CNN) algorithm. They used high-speed imaging data of two-phase flow in horizontal pipes to train the CNN model, which was able to accurately predict the flow pattern transition based on the input parameters such as phase superficial velocity and other geometrical features of the pipe.

In conclusion, the prediction of flow pattern in horizontal pipes is an important topic in the field of fluid mechanics. Several models and empirical correlations have been proposed to predict the flow regime transition, including Taitel and Dukler's [4] model (Fig. 2.5), the Baker [2] model, and the Lockhart-Martinelli [7] parameter-based model. In addition, machine learning approaches have been increasingly used to predict flow pattern in two-phase flow. The accurate prediction of flow pattern is crucial for designing and optimizing the performance of pipelines and other engineering applications.

2.4 Separation Technologies

The conventional vessel-type separation technology has been the go-to method for the petroleum industry. However, the conventional dividers are massive and costly. This drawback becomes more severe in offshore developments where platform costs are continuously increasing. To overcome these limitations, the industry is exploring alternative separator technologies, such as the Gas-Liquid Cylindrical

Cyclone (GLCC). The GLCC is a simple, compact, low-cost, and small inventory separator that can significantly reduce the size and cost of the separators. Research shows that phase separation is possible in simple T-junction, which can partially separate the phases and thus eliminate the need for a large separator. The development and application of such alternative separator technologies have become necessary due to the increasing cost of operation and the demand to implement such technologies at easy-to-operate smaller setups while fulfilling the demand.

2.5 Two-Phase Flow modeling

Initially, gas pipelines were designed with an efficiency factor in the flow equations to account for the pressure drop caused by the presence of a second phase. However, this approach was limited, and empirical correlations were later used to extend the single-phase pressure gradient equation for two-phase flows. These correlations, such as the Homogeneous Equilibrium Model (HEM), Separated Flow Model (SFM), and Drift-Flux Model, simplified the problem but were not suitable for analyzing different flow patterns. To address this issue, the two-fluid model was developed, which considers the phases separately using two sets of conservation equations and allows for the analysis of transients. While the flow mixture models are better suited for well-mixed dispersed flows, the two-fluid model is necessary for analyzing separated flows and is the preferred modeling approach. As an expert in Two Phase Flow and Technical Writing, it is important to understand the limitations of different modeling approaches and to choose the appropriate approach based on the specific characteristics of the flow being analyzed.

2.5.1 *Homogeneous Equilibrium Model (HEM)*

The Homogeneous Equilibrium Model (HEM) simplifies the analysis of two-phase flow by treating the gas and liquid phases as if they were combined into a single, well-mixed, pseudo-single phase with gas–liquid mixture properties. The model is a unified pressure derivative equation [8, 9].

$$-\frac{dp}{dx} = -\left(\tau_w \frac{\pi D}{A}\right)_{mix} - \rho_{mix} g \sin\theta + \rho_{mix} v_{mix} \frac{dv_{mix}}{dx} \quad (2.3)$$

This model assumes that the phases are in thermodynamic equilibrium and that they move at the same velocity (no-slip condition), whereas the density of the mixture can be calculated by

$$\rho_{mix} = A_G\rho_G + A_L\rho_L$$
$$\rho_{mix} = \alpha_G\rho_G + \alpha_L\rho_L \tag{2.4}$$

whereas the viscosity of the mixture can be given by

$$\mu_{mix} = A_G\rho_G + A_L\rho_L \tag{2.5}$$

The model uses a modified single-phase pressure gradient equation to describe the flow of the pseudo single phase. The fluid properties of the mixture are calculated as a volume fraction-dependent average of both phases. The gas–liquid mixture density is defined as a combination of the densities of the gas and liquid phases, while the gas–liquid mixture viscosity is a combination of the viscosities of the gas and liquid phases. These modifications are applied to the pressure gradient equation, resulting in an equation that describes the pressure gradient of the gas and liquid phases in the mixture. Similar modifications are made to the energy equation. As an expert in Two Phase Flow and Technical Writing, it is important to understand the assumptions and limitations of the Homogeneous Equilibrium Model and to use it appropriately based on the characteristics of the flow being analyzed.

2.5.2 *Lockhart and Martinelli Model*

The separated flow model, also known as the Lockhart and Martinelli Model, views two-phase flow as the flow of each phase in separate channels within the conduit, utilizing the hydraulic diameter concept. This model is commonly used to estimate frictional pressure drops in horizontal pipes if each phase moves independently of the other. The pressure drop in two-phase flow is calculated by multiplying the phasic superficial pressure gradient with a two-phase flow multiplier. While this model is limited to horizontal pipes, it provides a valuable framework for calculating pressure drops in many two-phase flow systems, including pipelines and heat exchangers. This model has been widely used in the oil and gas industry for designing and optimizing multiphase-flow systems. Phase superficial pressure change is multiplied by two-phase multiplier for frictional pressure gradient calculations [8].

$$\left(\frac{dp}{dx}\right)_f = \phi^2 L\left(\frac{dp}{dx}\right)_{SL} = \phi^2 G\left(\frac{dp}{dx}\right)_{SG} \tag{2.6}$$

$$\left(\frac{dp}{dx}\right)_f = \left(\frac{dp}{dx}\right)_k = \phi^2 k\left(\frac{dp}{dx}\right)_{Sk} \tag{2.7}$$

whereas

$\left(\frac{dp}{dx}\right)_{Sk}$ is the pressure gradient of the phase

$\left(\frac{dp}{dx}\right)_k$ represents pressure drop in kth phase

$\phi_G^2 = 1 + CX + X^2$ (dimensionless parameter for gas phase)
$\phi_L^2 = 1 + \frac{C}{X} + \frac{1}{X^2}$ (dimensionless parameter for liquid)
$X = \frac{\phi_G}{\phi_L}$ dimensionless parameter also known as Lockhart and Martinelli dimensionless parameter)

Empirical data is used to establish correlations for a parameter called X in two-phase flow. The constant C varies according to the characteristics of the fluids involved, with a value of 12 being used for situations involving laminar liquid flow and turbulent gas flow. The holdup, which is the proportion of a phase present in a mixture, is determined using an appropriate correlation. Specifically, the liquid holdup can be calculated using the Eq. 1 minus the quantity of 1 plus X raised to the power of 0.8 and then to the power of -0.378, as described by Chisholm [10]. Additionally, the value of another parameter called φ_k is determined based on the phasic hydraulic diameter, which is a measure of the flow area available for each phase. It is crucial to be familiar with the methods for computing holdup and the varying values of the constant C in different flow scenarios, as well as the role of φk, in order to accurately model and analyze two-phase flow systems.

2.5.3 Drift-Flux Model

The drift-flux model considers two-phase flow as a homogeneous mixture, assuming a correlation-dependent or constant difference in velocity between the phases based on local flow conditions. This allows for a better flow pattern-dependent modeling, considering the void fraction and the velocity difference between the phases. TACITE, a code developed for the analysis of transients in multiphase pipeline networks, uses separate continuity equations for gas and liquid phases, as well as mixture momentum and energy equations. Steady-state closure relationships based on flow pattern are employed to account for the missing momentum information. The model allows for flow regime transitions based on the continuity of variables, ensuring identical solutions for different flow regimes through the transition region. Pauchon et al. [11] conducted a study on the two-phase flow capabilities of TACITE, validating its effectiveness in analyzing transients.

2.5.4 Two-Fluid Model

A thorough method for resolving the two-phase flow problem is the two-fluid model. As a result, each phase can have its own velocity, pressure (or temperature) field, and mass and heat transfer terms between them. This is done by using distinct conservation equations in the Eulerian frame for each phase. The degree of coupling between phases is based on the model's assumption of interpenetrating continua. For mass transfer, a kinetic model must be used because strongly coupled phases frequently exhibit mechanical, thermal, but not always chemical, equilibrium [12].

The ability of the two-fluid model to account for transient and non-equilibrium conditions by describing the phasic pressure and temperature fields makes it applicable to a variety of flow conditions and scales, from smaller-scale one-dimensional pipe flow analysis using Euler equations to larger scale one-dimensional Navier–Stokes equation-based analysis of particle-scale flow. The model does, however, have limitations, such as conceptual challenges for specific flow conditions and the necessity of averaging methods for determining governing equations, which result in new unknowns that call for appropriate constitutive equations to close the equation set.

Utilizing additional field equations, such as a third mass and/or momentum balance equation, the model can be expanded to account for multiple phases and entrained liquid droplets, or gas bubbles. The two-fluid model is frequently used in the nuclear and petroleum industries; to analyze different pipe systems, the nuclear industry uses an eight-equation version, while the petroleum industry uses a seven-equation version [13].

2.5.5 Marching Models

Marching algorithms are numerical techniques that propagate the solution incrementally from the inlet to the outlet to resolve pressure drop calculations over short lengths of pipe. There are three main reasons why this tactic, referred to as "marching," is required.

- First, marching algorithms enable the application of mechanistic modeling techniques that make use of distinct flow models. For instance, after determining the flow pattern for the following pipe increment, the correct model is then applied to calculate the pressure drop.
- Second, marching algorithms have much lower computational costs than approaches that solve multiple points in a pipe simultaneously. This is crucial for long transportation pipelines because storage can be a problem.
- Last but not least, marching algorithms allow for a more accurate tracing of flow pattern transitions because the solution advances in incremental steps. Longer computation times are the price for this, though.

A common model for marching algorithms is the one-dimensional, steady-state, two-fluid model [14]. Finite difference methods (FDM) are used to organize the governing partial differential equations (Euler equations) into ordinary differential equations (ODEs), which are then numerically integrated over small pipe increments.

Runge–Kutta (RK) numerical procedures are better suited for space marching algorithms than typical FDM-based methods, which obtain solutions for all discrete points simultaneously. In RK numerical computation, the governing equations are rearranged in the explicit dU/dx ODE form.

$$\frac{dU}{dx} + \frac{dF}{dx} = C \tag{2.8}$$

Here, U is a variable which can be represented as

$$U = f(\rho, v_G, T) \tag{2.9}$$

Therefore,

$$\frac{d}{dx}\begin{bmatrix} p \\ v_G \\ T \end{bmatrix} = \left[\begin{pmatrix} v_G\left(\frac{\partial \rho_G}{\partial p}\right)_T & \rho_G & v_G\left(\frac{\partial \rho_G}{\partial p}\right)_P \\ 1 & \rho_G v_G & 0 \\ \rho_G v_G \eta_G \left(c_p\right)_G & \rho_G v_G^2 & \rho_G v_G \left(c_p\right)_G \end{pmatrix}\right]^{-1} \begin{bmatrix} 0 \\ -F_{fG} - F_{gG} \\ Q_G \end{bmatrix} \tag{2.10}$$

2.6 Flow in Junctions

In a pipe network, it is crucial to have high phase separation. The easiest and effective solution is to separate gas and liquid phase using junctions. There are abundant technologies of the junctions, and the simplest one is a "T-junction." For single-phase flows, engineers have enough knowledge to design the downstream equipment efficiently by predicting the flow split. It is even more problematic that a diversion exists in the junction, since it is very complicated to judge which phase would divert in the side arm and what would be the proportion.

Two main aspects of flow diversion in the junction are discussed here. Firstly, it is ideal that both phases will divide into two streams such as liquid and gas. Moreover, in practical situations, the attainment of discrete phases in every branch of a T-junction is frequently unattainable; instead, an asymmetrical distribution among the branches is more prevalent. The partitioning of multiphase flows at T-junctions entails an intricate interaction of multiple variables in contrast to the splitting of single-phase flows. Consequently, a comprehensive inquiry has been undertaken to improve our understanding and predictive precision of phase distribution.

Failure to comprehend these phenomena can result in suboptimal designs of subsequent equipment, leading to diminished operational efficiency. Figure 2.6 presents a practical illustration, depicting the isometric pipe layout of an industrial facility located in Kuwait. In the given configuration, the substances extracted from a low-pressure separator are channeled into two storage tanks, wherein a noticeable disparity has been noted. Tank 1 consistently maintained a high level of capacity, whereas Tank 2 consistently experienced underutilization.

In industrial settings, it is crucial to have a thorough understanding of the behavior of gas–liquid flow at junctions, as illustrated in Fig. 2.6 [15]. The flow distribution at a T-junction can be significantly affected by the arrangement of pipework, as

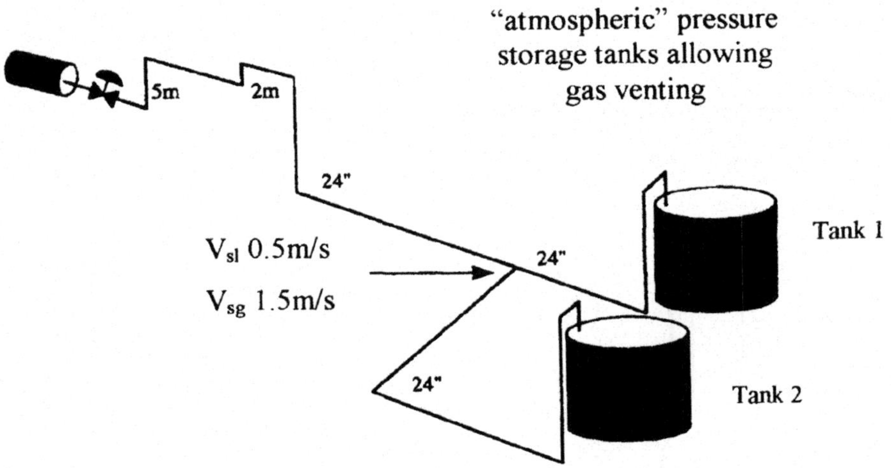

Fig. 2.6 Flow separation in the industries [15]

demonstrated in the case study of an industrial plant in Kuwait. The imbalanced distribution of fluid flow at the T-junction originated from disparities in the resistances downstream, leading to significant challenges related to slug flow within the system. In order to tackle this particular challenge, the configuration of the pipework was modified in order to achieve a more equitable distribution of fluid between the two storage tanks. This adjustment effectively alleviated the vibration issues that were attributed to the occurrence of slugging.

In contrast, T-junctions have been shown to function as partial phase separators, as evidenced by the research conducted by Oranje [16]. Oranje [16] proposed empirical guidelines for the prediction of phase distribution at T-junctions, which are based on the characteristics of the two phases within the pipe and the geometric properties of the downstream system. Nevertheless, the aforementioned guidelines were subject to certain limitations imposed by specific boundary conditions and underlying assumptions. This highlights the importance of comprehending the complex dynamics of gas–liquid flows at T-junctions, not only to enhance operational efficiency but also to achieve precise forecasting and optimal design.

References

1. E. Wren, B.J. Azzopardi, Affecting the phase split at a large diameter T-junction by using baffles. Exp. Therm. Fluid Sci. **28**, 835–841 (2004). https://doi.org/10.1016/j.expthermflusci.2003.12.017
2. O. Baker, Design of pipelines for the simultaneous flow of oil and gas, in *SPE Annual Technical Conference and Exhibition?* (1953). https://doi.org/10.2523/323-g

3. B.A. Eaton, C.R. Knowles, I.H. Silberbrg, The prediction of flow patterns, liquid holdup and pressure losses occurring during continuous two-phase flow in horizontal pipelines. J. Pet. Technol. **19**, 815–828 (1967). https://doi.org/10.2118/1525-pa
4. Y. Taitel, A.E. Dukler, A model for predicting flow regime transitions in horizontal and near horizontal gas-liquid flow. AIChE J. **22**, 47–55 (1976). https://doi.org/10.1002/aic.690220105
5. J. Mandhane, G.A. Gregory, K. Aziz, A flow pattern map for gas-liquid flow in horizontal pipes. Int. J. Multiph. Flow **1**, 537–553 (1974)
6. J. Zhang, X. Wei, Z. Wang, The recognition algorithm of two-phase flow patterns based on GoogLeNet+5 coord attention. Micromachines **14** (2023). https://doi.org/10.3390/mi14020462
7. R.W. Lockhart, R.C. Martinelli, Proposed correlation of data for isothermal two-phase, two-component flow in pipes. Chem. Eng. Prog. 38–48 (1949)
8. C.T. Crowe, F. Group, *Multiphase Flow Handbook*. (2006)
9. O. Shoham, *Mechanistic Modeling of Gas-Liquid Two-Phase Flow in Pipes*. (2006). https://api.semanticscholar.org/CorpusID:210916162
10. D. Chisholm, A theoretical basis for the Lockhart-Martinelli correlation for two-phase flow. Int. J. Heat Mass Transf. **10**, 1767–1778 (1967). https://doi.org/10.1016/0017-9310(67)90047-6
11. C.L. Pauchon, H. Dhulesia, G.B. Cirlot, J. Fabre, TACITE: a transient tool for multiphase pipeline and well simulation, in *SPE Annual Technical Conference and Exhibition?* https://doi.org/10.2118/28545-MS
12. M. Ishi, T. Habiki, *ThermoFluids dynamics of Two Phase Flow*. (2000)
13. C. Frepoli, J.H. Mahaffy, K. Ohkawa, Notes on the implementation of a fully-implicit numerical scheme for a two-phase three-field flow model. Nucl. Eng. Des. **225**, 191–217 (2003). https://doi.org/10.1016/S0029-5493(03)00159-6
14. L.F. Ayala, D. Alp, Evaluation of "Marching Algorithms" in the analysis of multiphase flow in natural gas pipelines. J. Energy Resour. Technol. **130** (2008). https://doi.org/10.1115/1.3000103
15. E. Wren, Geometric effects on phase split at a large diameter T-junction. Phd. Diss., (2001)
16. L. Oranje, Condensate behavior in gas pipelines is predictable. Oil Gas J. **71**, 39–44 (1973)

Chapter 3
Effect of Parameters on Phase Separation

This book plays a pivotal role in establishing a fundamental basis, laying the groundwork for subsequent inquiries aimed at improving the geometric properties of T-junctions. In recent years, extensive research endeavors, conducted between 2006 and 2020, have been dedicated to enhancing the performance of T-junction geometry. This chapter discusses in detail the effect of diameter ratio (DR), velocity ratio (VR), side arm inclination on phase separation as well as state of the art converging and diverging T-junction designs to achieve high separation efficiencies.

In a study undertaken in 2006, Yang et al. [1] investigated multiple aspects pertaining to the enhancement of phase separation in T-junctions. A study conducted by Kharkeshi et al. [2] focused on the dissipation of slug flow, a flow regime that has gained prominence in various industrial applications. The present study provides a comprehensive analysis of slug flow, elucidating its distinctive features and behavioral patterns, thereby contributing significant knowledge regarding its dynamics within T-junctions. This study emphasized the importance of comprehending and effectively managing slug flow in industrial settings, highlighting its significance in optimizing the overall performance of the system.

3.1 Types of T-Junction

Following are the types of T-junctions based on flow direction as well as the diameter ratio.

F. Ejaz et al., *Two Phase Separation in the T-Junction*,
https://doi.org/10.1007/978-981-99-9354-3_3

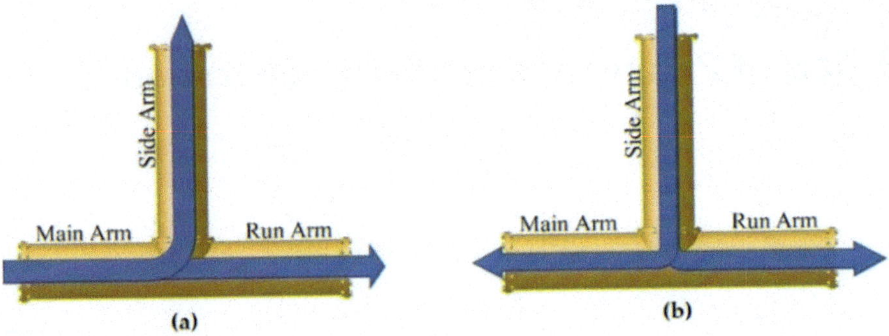

Fig. 3.1 Categories of T-junctions; **a** Branching; **b** Impacting

3.1.1 Classification Based on Flow Direction

Two distinct types of T-junctions can be classified based on the direction of flow: branching T-junctions and impacting T-junctions [3]. The visual representation of these types is depicted in Fig. 3.1.

(a) **Branching T-Junction**

Within a branching T-junction configuration, the predominant fluid flow traverses the main arm before subsequently bifurcating into two distinct pathways, specifically the run arm and side arm. Flow separation is observed to take place precisely at the midpoint of the junction, resulting in the division of the stream.

(b) **Impacting T-Junction**

The term "impacting T-junction" is used to describe the behavior of the fluid stream in this particular junction. In this scenario, the fluid enters the main pipe through a side arm and undergoes a forceful impact against the pipe wall. As a consequence, the stream is divided into two separate flows within the pipe: the main arm and the run arm. Figure 3.1 presents a schematic depiction that serves as a visual aid to facilitate comprehension of this particular configuration.

3.1.2 Classification Based on the Ratio of Side Arm Diameter

The categorization of T-junctions can also be predicated on the ratio of diameters involved. Figure 3.2 depicts the various categories delineated by these ratios.

A T-junction is classified as a regular T-junction when the diameter of the side arm is equal to the diameter of the main arm. Conversely, when the diameter of the side arm is smaller than that of the main arm, it is classified as a diverging T-junction.

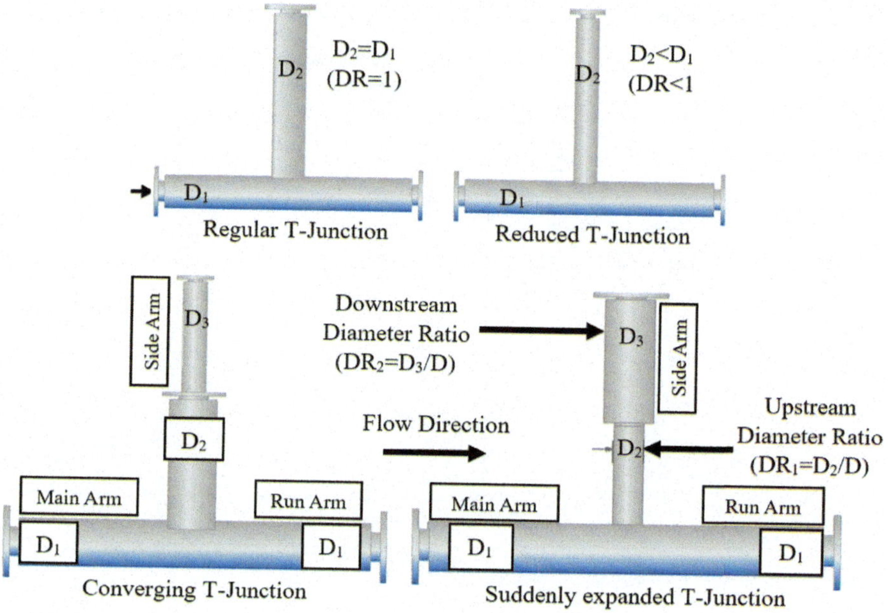

Fig. 3.2 Types of T-junction w.r.t side arm diameter ratio

In the scenario of a T-junction where the flow converges and then abruptly expands, the diameters of the side arms are represented as DR_1 and DR_2. These values correspond to the ratios of the diameters of the upstream and downstream pipes in the side arm, respectively. Moreover, in order to sustain the abruptly changing side arm, DR_2 consistently exhibits a higher value compared to DR_1.

3.2 Criteria to Represent Phase Redistribution

The criterion to depict the phase redistribution phenomenon, graphically, is illustrated in Fig. 3.3. Lower half of the plot implies the gas-dominated region and upper half as the liquid-dominated. The equal split line indicates no phase separation with liquid and gas sharing equal volume in the side arm outlet. Equations (3.1) and (3.2) present the formulae to calculate the fraction of gas and liquid in the side arm. Whereas "W_G" and "W_L" represent the mass flow rate of gas and liquid, respectively. The subscripts 1 and 3 are for main arm and side arm. To evaluate the separation performance of the T-junction, two types of methods are surveyed from the literature. Comparatively, the most efficient method is to calculate the separation efficiency using the area under the curve. More details on the justifications of selecting this method were mentioned by Pao and Memon [4] and will not be repeated here.

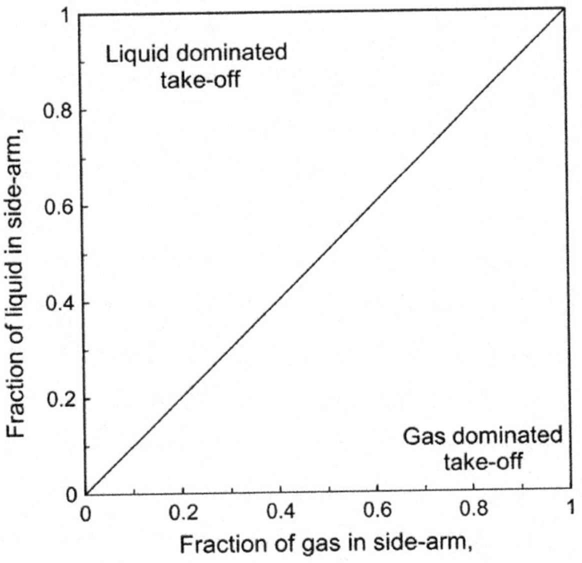

Fig. 3.3 Phase separation data representation

$$F_G = \frac{W_{G3}}{W_{G1}} \tag{3.1}$$

$$F_L = \frac{W_{L3}}{W_{L1}} \tag{3.2}$$

Comparatively, the most efficient method is to calculate the separation efficiency using the area under the curve. This method of calculating separation efficiency was successfully implemented by Pao and Memon [4]. Separation efficiency can be computed by evaluating area under the curve. This method operates such that 0% separation efficiency can be assigned to the equal split line and 100% separation efficiency for horizontal axis with no liquid carryovers. Efficiency determined by this method can be used as a key performance index of the efficiency of a particular T-junction design.

3.3 A Unique Separation Criteria

The evaluation of the effectiveness of a T-junction and the comparison of various geometric configurations can be efficiently performed by employing metrics such as Peak Liquid Carryover (PLC) and Liquid Carry-Under Threshold (LCT). The dimensionless parameter denoted as "S" was introduced by Minh et al. [5] in order to assess the effectiveness of separation. In the pursuit of attaining optimal phase separation, it is advantageous to possess a low value for the phase lower critical solution temperature (PLC) and a high value for the liquid–liquid critical temperature (LCT). Minh et al. presented a case in which a discrepancy arises between two T-junctions, one exhibiting a high LCT/PLC ratio and the other displaying a lower PLC/LCT ratio. In order to tackle this issue, the researchers put forth a consolidated parameter denoted as "S" that serves as a quantitative measure for evaluating the performance of T-junctions. The computation of variable "S" entails the determination of the integral of a particular function.

The parameter "S" exhibits a positive correlation with the velocity ratio, as depicted in Fig. 3.4. In order to achieve optimal phase separation, it is imperative to minimize the value of the parameter denoted as "S." This suggests that modifying variables such as augmenting the liquid superficial velocity or diminishing the superficial gas velocity will result in a decrease in the area beneath the curve, leading to a diminished "S" value. The aforementioned observation aligns with the results reported in the extant body of literature. The study conducted by Minh et al. utilized the Volume of Fraction (VOF) method in combination with the k-ε turbulence model.

Minh et al. [5] study conducted a comparative analysis of performance between reduced and regular T-junction configurations. The investigation conducted by the researchers demonstrated that a T-junction with a reduced diameter ratio of 0.6 exhibited decreased PLC values and increased LCT values when compared to a standard T-junction. This discovery aligns with previous research results, emphasizing the importance of PLC and LCT as crucial measures for accurately assessing and enhancing the effectiveness of T-junction configurations for phase separation operations.

The predominant area of study has predominantly concentrated on T-junctions where the main arm is oriented in a horizontal manner. The annular and stratified flow regimes have received significant attention in research, primarily because of their widespread occurrence. The phenomenon of slug flow, despite its negative impact on equipment, has also been the focus of extensive research. Wren [6] discussed a

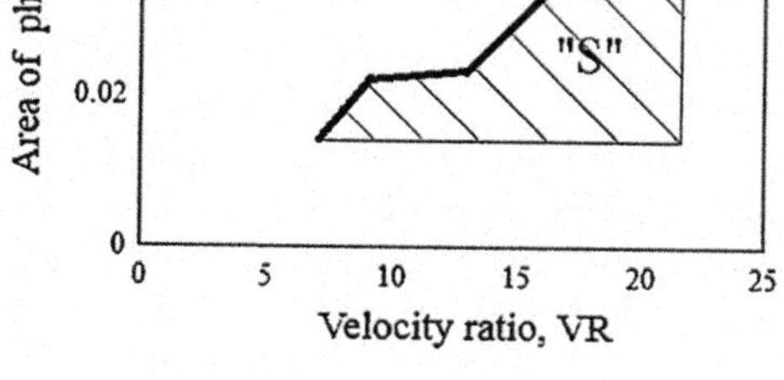

Fig. 3.4 New criteria for separation performance [5]

novel approach wherein they suggested the installation of a T-junction featuring a vertically upward branch arm within the main pipeline situated on the seabed. The main objective of this concept was to design a separator that could effectively manage slugs caused by diverse topography, consequently mitigating the requirement for expensive onshore slug catchers. In further elaboration of this notion, Katsaounis and Schultheiss [7] developed dynamic separator systems with the objective of mitigating the occurrence of slugging in multiphase pipelines. Nevertheless, up until now, none of these suggested T-junction separators has been put into practice in real-world scenarios. The observed phenomenon can be ascribed to the notable influence of the inlet flow regime on the separation performance of T-junctions, a characteristic that tends to vary as the oil field undergoes maturation.

Oil companies frequently opt to depend on established and proven equipment due to the intricate technical intricacies involved. However, there has been a recent advancement in T-junction technology. At the BP/Amoco Chemicals facility in Hull, a T-junction is presently employed as a partial phase separator. The obstacle encountered pertained to the incomplete vaporization of the liquid product as it traversed a valve prior to entering the distillation column. The upward movement of the liquid, in conjunction with the vapor, within the column resulted in a decrease in column efficiency. The conventional approach entails the installation of a separator to redirect the residual liquid and vapor to distinct locations within the column. Nevertheless, this methodology is not only costly but also time-consuming and unfeasible within the context of an existing operational facility. By capitalizing on the understanding of phase separation phenomena at T-junctions, it was hypothesized that this technological approach could present an optimal solution. The feasibility of the proposed geometric arrangement was investigated by Azzopardi et al. [8].

In the present configuration, the biphasic flow is introduced into the system, and upon traversing the initial curvature, liquid droplets carried by the gas phase come into contact with the opposite wall. The liquid substance moves downward in the form of a thin layer, gathering in the U-shaped section, consequently guiding the gas with fewer impurities to travel along the secondary arm of the T-shaped intersection. Hence, the present configuration of a basic pipe system, featuring a T-junction at its core, functions as a partial phase separator, facilitating the extraction of a gas-enriched stream while retaining a liquid-enriched stream. After the installation process, it has been empirically demonstrated that the proposed geometric arrangement operates efficiently, effectively achieving the desired separation between the liquid and vapor streams. This configuration enables the gas-rich and liquid-rich streams to enter the column at more appropriate locations. As a result, the restoration of efficiency has been achieved with minimal cost or disruption, thereby prompting the examination of T-junctions as a potentially feasible substitute for traditional separators in subsequent endeavors.

The observed efficacy of the partial phase separation exhibited by the aforementioned T-junction configuration has the potential to serve as a catalyst for other enterprises to investigate the utilization of T-junction technology in addressing their unique obstacles. Hence, it is imperative to prioritize ongoing investigations concerning phase separation occurring at T-junctions. This book aims to enhance our comprehension by examining the impact of various physical geometries, such as side arm diameter, side arm orientation, and junction modifications, on the phenomenon of phase separation occurring at T-junctions. Given the extensive existing research on the factors influencing phase split at junctions, this discussion will solely reference pertinent literature. A number of comprehensive literature reviews have been published that extensively discuss the various aspects of T-junctions and phase separation. Notably, these reviews include works by Lahey (1987), Muller and Reinman (1991), Azzopardi and Hervieu (1994), and Azzopardi (1999).

3.4 Influence of Side Arm Inclination

Chen et al. [9] have demonstrated that the phase distribution is significantly influenced by angle between the arms. Flow regimes examined in their study encompassed slug flow, annular flow, and annular-slug flows. The main arm was held in a horizontal position, while the secondary arm was positioned at various angles between 30° and 150° relative to the main arm (see Fig. 3.5). In order to achieve precise observations of flow patterns, a high-speed camera was employed.

The data presented in Fig. 3.6a clearly demonstrate that an increase in the inclination angle leads to a noticeable increase in phase separation in slug flow. This phenomenon can be ascribed to the fluid's inclination to maintain its flow in the main pipe owing to inertia, thereby reducing the likelihood of it diverting into the secondary pipe. In contrast, Fig. 3.6b illustrates a reduced occurrence of phase separation in annular flow. The observed phenomenon can be attributed to the development of a secondary flow streamline along the opposing wall of the junction. This secondary flow induces the downward movement of the liquid into the side arm. Nevertheless, when the branch angles are increased, the presence of inertia effect poses a significant

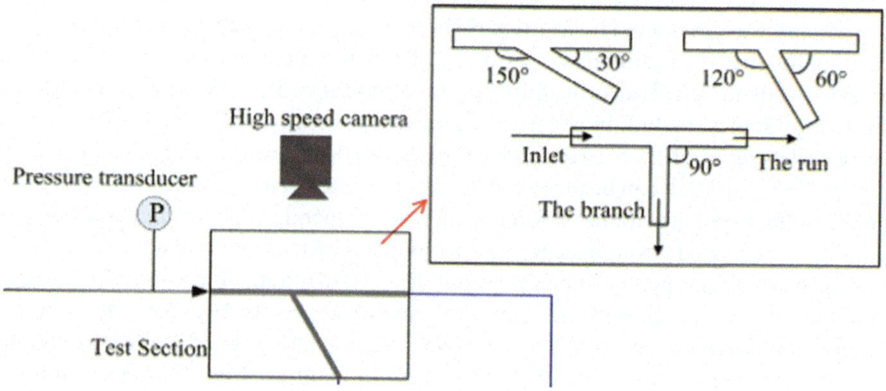

Fig. 3.5 Side arm inclinations used by Chen et al. [9]

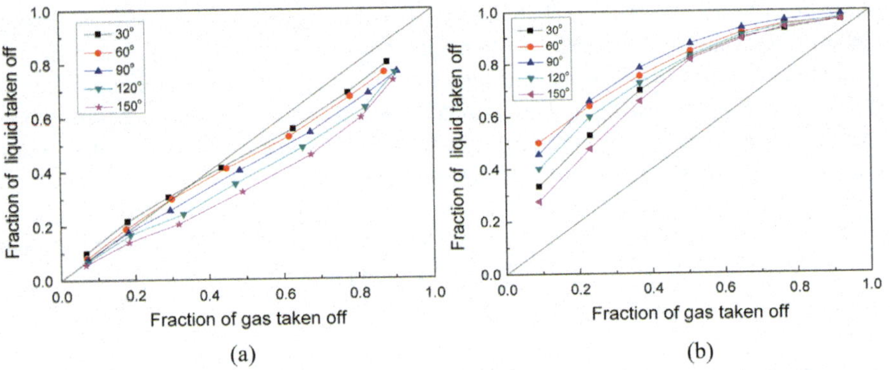

Fig. 3.6 Effect of side arm angle [9] **a** Slug Flow; **b** Annular Flow

obstacle for the liquid stream to alter its trajectory, thereby resulting in a decrease in the efficiency of separation.

In a separate investigation carried out by Hong [10], the impact of side arm inclination was examined, with angles spanning from +90° to −90°. The findings of his study indicated that when the side arm was inclined at a vertical angle of +90°, there was a notable reduction in the quantity of liquid redirected to the side arm. This decrease can be attributed to the combined influence of gravitational force and inertia, which caused the liquid to remain predominantly within the main arm. According to the data presented in Fig. 3.7, a decrease in the angle of inclination led to a corresponding increase in the percentage of liquid that was redirected toward the branch of the junction. The most significant impact was observed when the angle of inclination reached −90°, corresponding to the vertically downward side arm.

The influence of side arm orientation on phase separation is illustrated in Fig. 3.8 [10]. The splitting ratio for a side arm inclined at an upward angle of +20° is depicted in Fig. 3.8 [11]. The splitting ratio for downward side arms at angles of −40° and −60° is depicted in Fig. 3.9 [11].

A study was conducted by Marti and Shoham [11] to investigate the impact of side arm inclination on stratified wavy flow. A T-junction configuration was employed, consisting of a vertical run arm and a horizontal side arm. The side arm was positioned at various angles, including negative angles of −5°, −10°, −25°, and −40°, as well as positive angles of 1°, 5°, 10°, and 20° in the vertical upward direction. According to Figs. 3.8 and 3.9, it can be observed that the presence of a negative inclination leads to the influence of gravitational force, which subsequently causes a higher volume of liquid to be diverted through the branch in comparison to an upward inclination. In the scenario where the inclination is directed downwards, a notable rise in the amount of liquid carryover through the side arm was observed. In Fig. 3.8, the side arm inclination is depicted as +20°, resulting in a significant gas yield while maintaining liquid carryovers below the threshold of 0.9. In contrast, Fig. 3.9 demonstrates a significant rise in the proportion of liquid and a decrease in the proportion of gas within the side arm. Nevertheless, when considering positive inclinations, such as

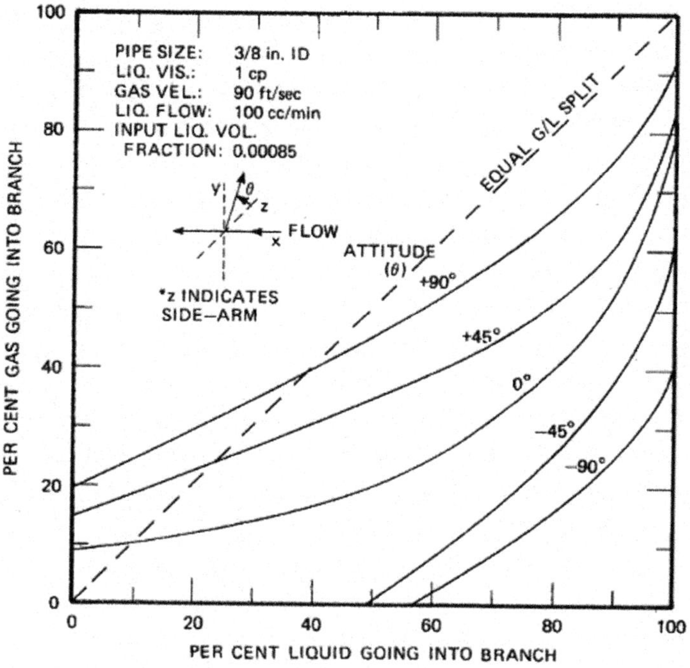

Fig. 3.7 Influence of side arm orientation on phase separation [10]

an upward side arm, the force of gravity supports the movement of liquid toward the run arm. Table 3.1 presents a comprehensive overview of the impact of side arm inclination on the process of phase separation.

3.5 The Impact of Superficial Velocities

The influence of gas and liquid superficial velocities on the process of phase separation is significant, as they have the ability to affect the dynamics of separation. When a fluid stream is in motion, its inertia and momentum play a crucial role in either hindering or facilitating the separation process. In an experimental study conducted by Al Azizi [16], it was observed that an increase in the liquid superficial velocity led to substantial separation efficiencies. Since the liquid superficial velocity increases, so does its momentum and inertia. The increased inertia of the liquid makes it less inclined to divert into the side arm, as illustrated in Fig. 3.10.

Figure 3.11 demonstrates the impact of liquid and gas superficial velocities on phase separation [16]. In Fig. 3.10a, the gas velocity (V_G) is measured to be 2.5 m per second. In Fig. 3.10b, the liquid velocity (V_L) is observed to be 0.09.

Figure 3.11 examines the impact of gas and liquid superficial velocities for two different cases: (a) when the liquid superficial velocity (VL) is 0.0055 m/s, and

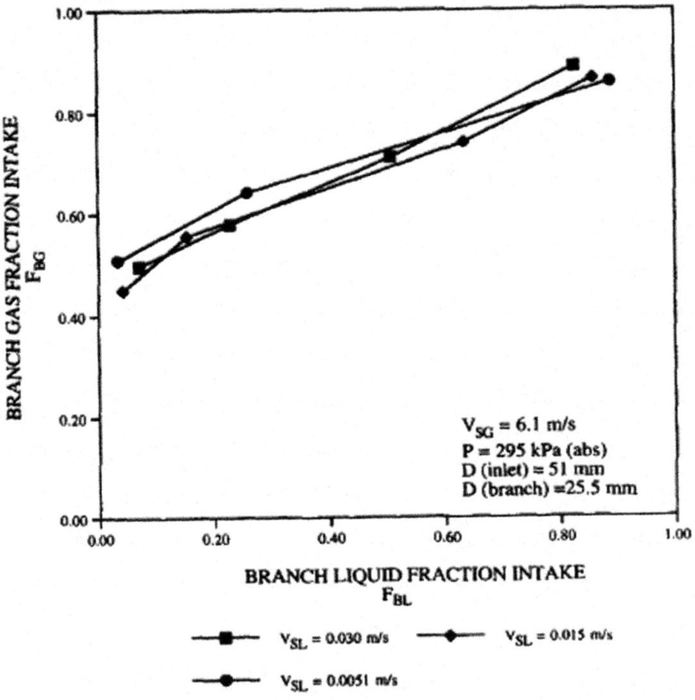

Fig. 3.8 Splitting ratio for +20° upward side arm [11]

(b) when the liquid superficial velocity (VL) is 0.0097 m/s [17]. Das et al. [17] conducted a study to investigate the impact of the superficial velocities of both gas and liquid on phase redistribution using a simple T-junction configuration. Based on the data presented in Fig. 3.11, an increase in gas superficial velocity results in a significant rise in the liquid fraction in the side arm. This can be attributed to the higher pressure drop experienced under these conditions. Conversely, at lower liquid superficial velocities, the data points are closer to or slightly above the equal split line. This indicates that at lower velocities, the liquid possesses less inertia, leading to its extraction from the side arm due to pressure drop. However, it was observed that higher liquid velocities improved phase separation. These findings align with the results reported by Rea and Azzopardi [18] and Peng et al. [19].

Saba and Lahey [20] conducted an experiment on a T-junction with a main pipe diameter of 0.0381 m, analyzing slug and bubbly flow regimes. They concluded that even at extremely low inlet qualities, significant phase separation could be achieved. The researchers also noted that variations in the inlet mass flux did not affect the process of phase separation.

Seeger et al. [14] thoroughly investigated the effect of liquid and gas superficial velocities on phase separation. They found that when the liquid superficial velocity was kept constant, there was a decline in phase separation as the gas superficial

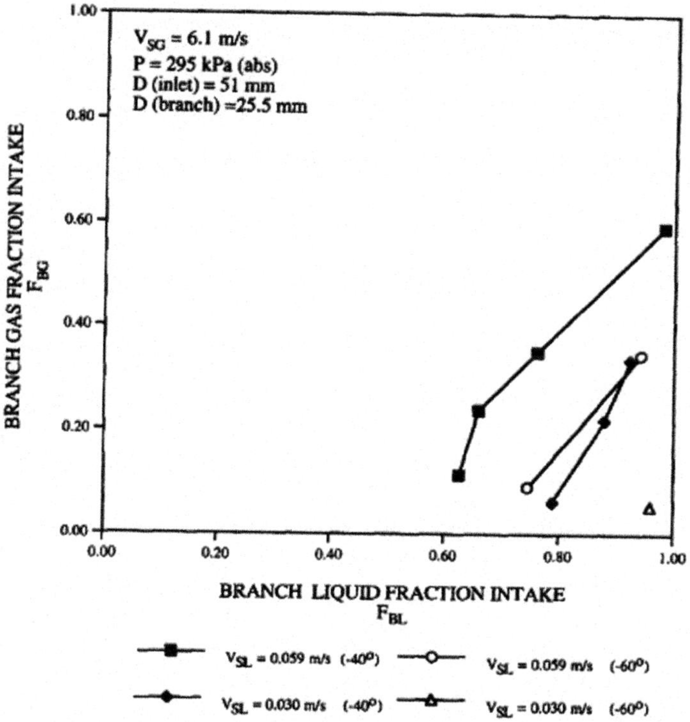

Fig. 3.9 Splitting ratio for −40 and −60° downward side arm [11]

velocity increased. The data were collected for a T-junction with a horizontal side arm. However, in the case of a downward side arm, an increase in phase separation was observed as the gas superficial velocity increased, although the effect was not particularly significant. Additionally, when the air superficial velocity was kept constant, an increase in the liquid superficial velocity resulted in an enhancement of phase separation. Reimann and Seeger [21] also noted that factors such as mass split ratio, geometric characteristics, and flow patterns influenced the phenomenon of phase separation in T-junctions. Azzopardi et al. [8] reported similar findings, indicating that an increase in liquid superficial velocity improved phase separation.

Rea and Azzopardi [22] investigated the phenomenon of phase separation in a regular T-junction. They used a T-junction with a larger diameter main arm (0.127 m) for comparative purposes. The gas and liquid superficial velocities employed in the study ranged from 0.014 to 0.558 m/s and 4 to 30 m/s, respectively. Their findings revealed that in T-junctions with larger diameters, a greater accumulation of liquid occurred in the lower section of the pipe. As a result, when the liquid superficial velocities were decreased, the occurrence of liquid carryover was minimal. Pandey et al. [23] conducted a study using water and kerosene. The team employed a T-junction with a main arm diameter of 0.025 m and varied superficial velocities ranging from

Table 3.1 Summary of literature (effect of side arm inclination on phase separation)

Authors	Side arm inclination (degrees)	DR	Working fluids	Gas superficial velocity (m/s)	Liquid superficial velocity (m/s)	Flow pattern
Penmatcha et al. [12]	0, 1, 5, 10, 20, 35, −5, −10, −25, −40, −60	1	Air–Water	6.1	0.0051–0.059	Stratified Wavy
Marti and Shoham [11]	0, 1, 5, 10, 20, 35, −5, −10, −25, −40, −60	0.5, 1	Air–Water	6.1	0.0051–0.059	Stratified
Azzopardi and Smith [13]	0, 90	0.67, 1	Air–Water	0.7–31	0.008–0.078	Stratified and Annular
Seeger et al. [14]	0, +90, −90	1	Air–water	10–40	0.5–4	Slug, Bubbly, Annular
Wren [6]	0, 90, −90	1	Air–Water	6.2–24	0.04–0.31	Stratified and Annular
Hong [10]	0, 45, 90	1	Air–Water	9.1–42.7	0.0023–0.47	Stratified Annular
Ottens et al. [15]	0, 0.1, 0.25, 0.5	1	Air–Water	5–14	0.0009–0.014	Stratified

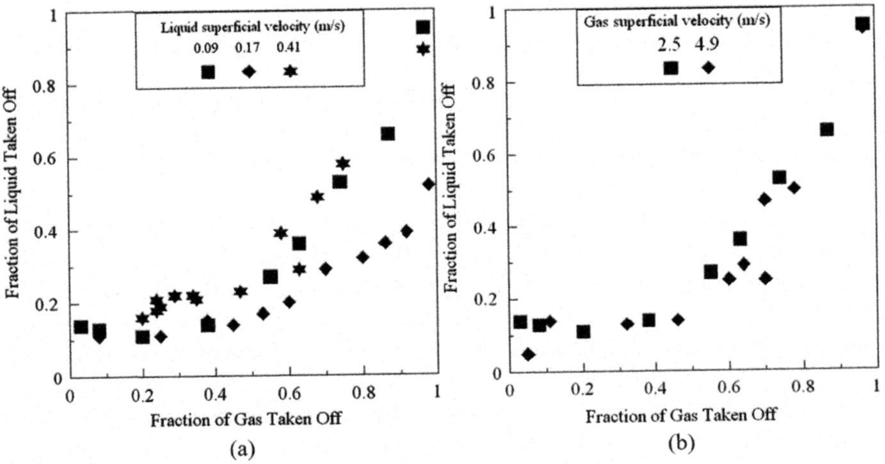

Fig. 3.10 Effect of liquid and gas superficial velocities on phase separation [16]; **a** $V_G = 2.5$ m/s; **b** $V_L = 0.09$ m/s

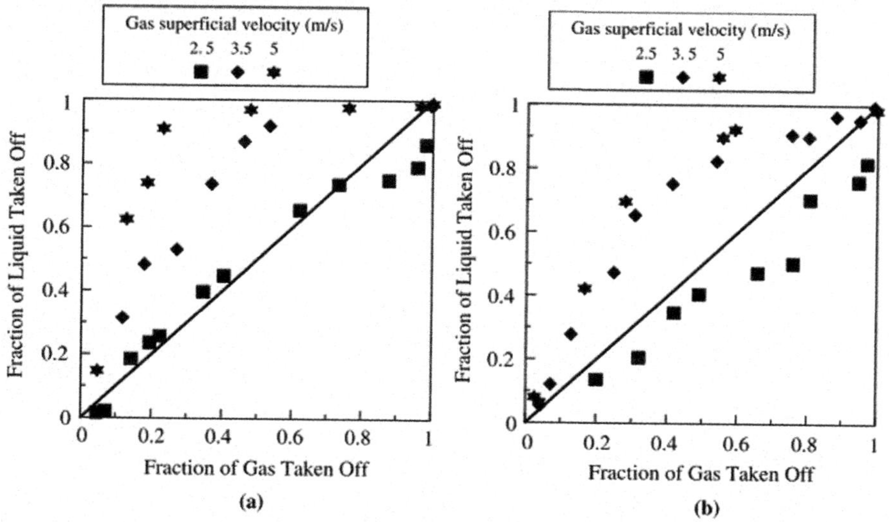

Fig. 3.11 Influence of gas superficial velocity and liquid superficial velocity of **a** $V_L = 0.0055$ m/s; **b** $V_L = 0.0097$ m/s [17]

0.05 to 2 m/s. The study generated three distinct flow patterns: stratified flow, plug flow, and stratified wavy flow. The results showed that when the kerosene superficial velocity was kept constant, an increase in water superficial velocity enhanced phase separation. Conversely, an increase in kerosene superficial velocity resulted in an escalation of liquid carryover.

Van Gorp [24] reported contradictory findings to those of Rea and Azzopardi [22]. They used a T-junction with a main arm diameter of 0.0381 m and a diameter ratio of 0.206. The observed velocities of the liquid and gas ranged from 0.0021 to 0.0398 m/s and from 2.7 to 40.1 m/s, respectively. Based on their findings, it was observed that increasing the gas superficial velocity did not significantly divert the gas phase into the side arm. The authors explained that elevated gas superficial velocities increased the momentum of the gas phase, making diversion to the side arm challenging under such conditions.

Figure 3.12 presents the impact of liquid (water) and gas (air) superficial velocities on the liquid carryover threshold (LCT) [25]. In Fig. 3.12a, the gas superficial velocity (V_G) is 0.228 m/s, while in Fig. 3.12b, the liquid superficial velocity (V_L) is 0.094 m/s.

Saieed et al. [25] introduced the concept of the "liquid carryover threshold (LCT)" and established a correlation between the liquid–liquid contact time (LCT) and the superficial velocities of both phases. The study examined superficial velocities for different diameter ratios to conduct a comprehensive analysis. When investigating the impact of liquid superficial velocity, a constant gas superficial velocity of 0.228 m/s was maintained. Similarly, when examining the influence of air superficial velocity, a constant water superficial velocity of 0.094 m/s was maintained. The data presented

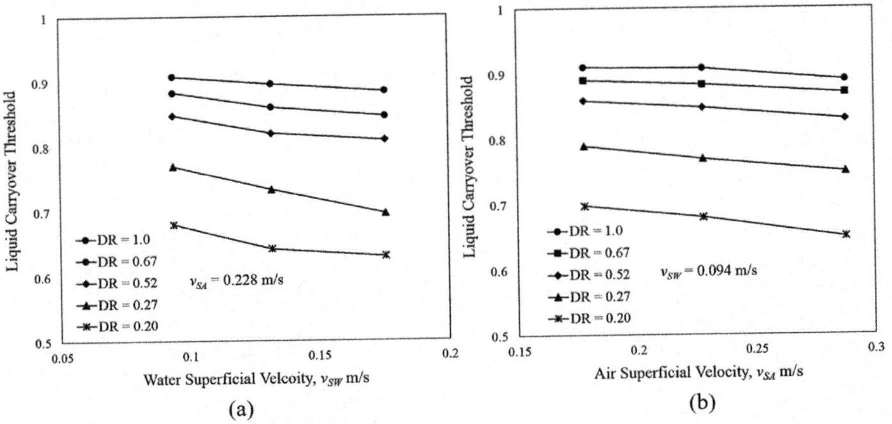

Fig. 3.12 Influence of liquid (water) and gas (air) superficial velocity on LCT [25]; **a** $V_G = 0.228$ m/s; **b** $V_L = 0.094$ m/s

in Fig. 3.12 shows that an increase in liquid superficial velocity leads to a decrease in the liquid carryover threshold (LCT). Furthermore, a reduction in the diameter ratio corresponds to a decrease in the overall liquid carryover threshold. The summary of relevant research articles is given in Table 3.2.

3.6 Effect of Diameter Ratio

Understanding the effect of diameter ratio, side arm inclination, and phase superficial velocities on phase separation is crucial in the context of the research on the novel suddenly expanded T-junction design. Extensive research available in the literature highlights the significance of these parameters. Therefore, it is important to discuss the relevant research studies related to the present work.

3.6.1 Origin of Inverse Relation (Year 1980–1990)

The investigation of the influence of diameter ratio on phase separation has been a focal point of scholarly inquiry since the 1980s. A study was conducted by Azzopardi and Whalley [28] to investigate the impact of flow patterns on T-junctions with diameter ratios of 0.2, 0.4, and 0.6. The authors placed particular emphasis on the correlation between phase separation and flow patterns, however, they did not specifically address the impact of the side arm diameter ratio. A subsequent study conducted by Azzopardi [29] aimed to enhance the existing research by investigating the impact of varying diameter ratios (0.4 and 1) on annular flow. The researcher's findings suggest

Table 3.2 Research studies on the effect of superficial velocities on separation efficiency

Researchers	Main arm diameter	Working fluids	Working pressure (bar)	Gas superficial velocity (m/s)	Liquid superficial velocity (m/s)	Flow pattern
Saba and Lahey [20]	0.038	Air–water	1.3–1.9	0.8–6.5	1.35–2.7	Stratified and Slug
Seeger et al. [14]	0.05	Air–water	0.7	10–40	0.5–4	Slug, Annular and Bubbly
Das et al. [17]	0.005	Air–water	1.3–1.4	1.5–15	0.05–0.09	Stratified
Wren et al. [26]	0.005	Air–water	1.5	1.14–4.56	0.093–0.313	Slug
Van Gorp [24]	0.038	Air–Water	3	2.7–40.1	0.0021–0.0398	Stratified, Stratified Wavy, Annular
Rea and Azzopardi [22]	0.127	Air–water	1	4–30	0.04–0.558	Stratified
Saieed et al. [27]	0.078	Air–water	1.01	0.178–0.288	0.094–0.76	Stratified Slug

that a decrease in the diameter ratio has a significant impact on phase separation, as depicted in Fig. 3.13.

The authors Shoham et al. [30] also emphasized the influence of the flow regime on the liquid phase behavior observed at the junction. In the context of fluid dynamics,

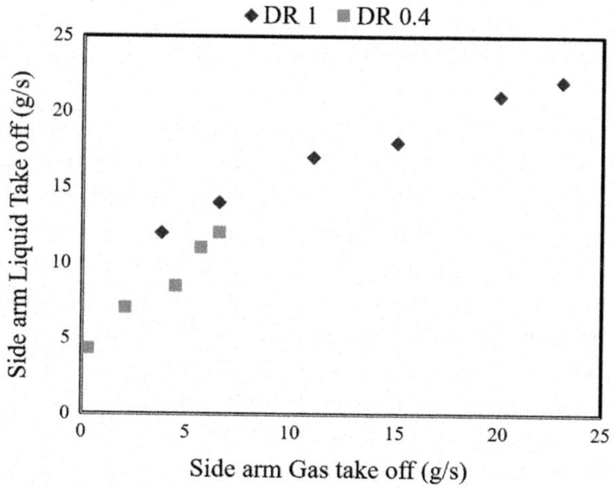

Fig. 3.13 Gas and liquid take off at DR 1 and 0.4 gas velocity = 55.6 kg/m^{2}s, liquid velocity = 79.4 kg/m^{2}s. Data Source: [29]

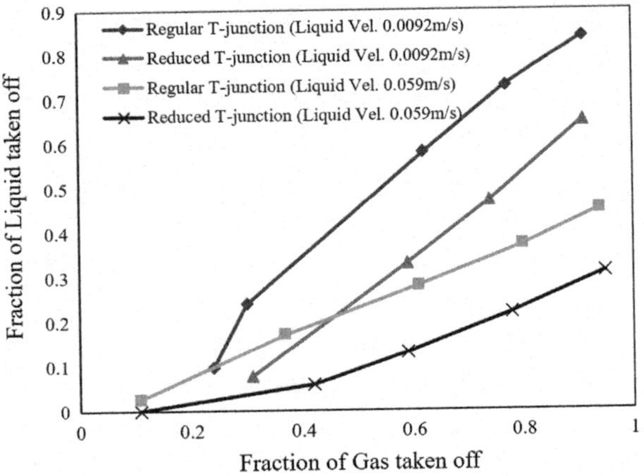

Fig. 3.14 Gas and liquid take off in stratified wavy flow (gas velocity = 6.1 m/s). Data Source: [31]

annular flow is characterized by the dominance of centripetal forces. On the other hand, the behavior of stratified wavy flow is contingent upon the specific flow rates involved. In the case of stratified smooth flow, the liquid typically exhibits a tendency to remain within the main arm of the flow. Expanding upon this notion, Shoham et al. [31] conducted an additional study for examining impact of separation performance improvement with the diameter ratio within identical flow conditions. A distinct impact of DR was observed when regular T-junction was compared. Results presented in Fig. 3.14 depict the overlay of two distinct liquid superficial velocities obtained from the conducted study. The findings demonstrate that reduced T-junctions exhibit a reduced amount of liquid take-off when compared to the same liquid and gas superficial velocities. Therefore, it can be inferred that a substantial reduction in the diameter ratio leads to a significant improvement in separation performance. Azzopardi [29] and Hwang et al. [32] had the similar findings.

3.6.2 *Agreements with the Inverse Relation (Year 1991–2000)*

Hart et al. [33] investigated the separation performance in reduced as well as conventional junction. The T-junction with a diameter 0.051 m and a modified T-junction of side arm of 0.038 m were employed. The study revealed that the reduced T-junction exhibited a notably higher minimum gas superficial velocity necessary for liquid ingress into the side arm, in comparison to the regular T-junction. The explanation provided supposed that the pressure difference resulting from the pressure drop in the side arm is inadequate, particularly when gas superficial velocities are lower and velocities of liquid phase are higher, for overcoming the threshold. Nevertheless, as soon adequate pressure drop is attained, liquid phase initiates toward the side branch. Additionally, it can be seen, when the liquid is in axial motion within the main arm, it can be elevated into the side arm by just a minimal pressure-drop. Nevertheless, when liquid possesses significant momentum, the inherent resistance to change in its motion necessitates the application of substantial pressure in order to displace it toward the side arm. The observed behavior suggests that a decrease in the diameter ratio has a positive impact on phase separation, particularly when extraction rates are lower.

Azzopardi and Smith [13] raised concern regarding the theoretical model put forth by Hart et al. [33], asserting that it exhibited a lack of alignment with empirical evidence. Nevertheless, the researchers deemed the prediction model to be acceptable for stratified wavy flow. The research conducted by Azzopardi and Smith [13] focused on the study of two-phase flow within a reduced T-junction configuration. Main arm had a diameter of 1.5 inches, while the side arm had a diameter of 1 inch. It was observed that annular flow exhibited higher liquid take-off at lower liquid inlet volume fractions. Within the context of annular flow, it is observed that the liquid film adheres closely to the inner wall and undergoes centripetal forces, potentially leading to increased liquid take-offs in comparison to alternative flow regimes.

The behavior of an industrial-scale T-junction of 0.43 m diameter was examined by Mudde et al. [34] across bubbly, stratified, and stratified wavy flow regimes. The researchers established a correlation between the behavior exhibited by a T-junction with a large diameter and that observed in a T-junction of smaller scale. The researchers have made a discovery regarding the significant influence of flow pattern in downstream or run arm. A real-time pulsating phenomenon was observed, wherein the presence of pulsating churn flow in the run arm had an impact on the phase behaviors. In cases where the pulsating effect is not readily observable, it remains feasible to estimate the pressure-drop in the junction by employing experimental techniques or theoretical models.

The experimental investigations on a reduced T-junction were conducted by Griston et al. [35]. An adjustable device, referred to as a spliTigator, as depicted in Fig. 3.15, was employed to regulate the phase split behavior occurring at the side arm. The dimensions of the side arms employed in the reduced T-junction experiment were 3×2 inches and 4×2 inches. The obtained data was subsequently compared with other T-junctions. Figure 3.16 illustrates a noticeable decrease in the liquid take-off to the side arm when employing the reduced side arm configuration. The reduced aperture size of the side arm presents a greater difficulty for the liquid to initiate flow into the side arm when velocities are lower. According to Fig. 3.17, it can be observed that the reduced T-junction exhibits favorable performance in relation to phase separation. The impact of incorporating the spliTigator into the reduced T-junction is limited to the phase separation process occurring at inlet qualities exceeding 70%.

The impact of side arm angle on phase separation was examined by Marti and Shoham [11]. The researchers conducted a comparative analysis of the findings, as depicted in Fig. 3.18. The performance of the reduced T-junction was superior, particularly at mass split ratios ranging from low to intermediate levels. Further investigation is required to examine the impact of a slight upward inclination of the side arm (specifically, at a 5-degree angle) on phase separation, particularly in relation to its potential contribution to excessive liquid take-off, especially when aiming for higher extraction ratios.

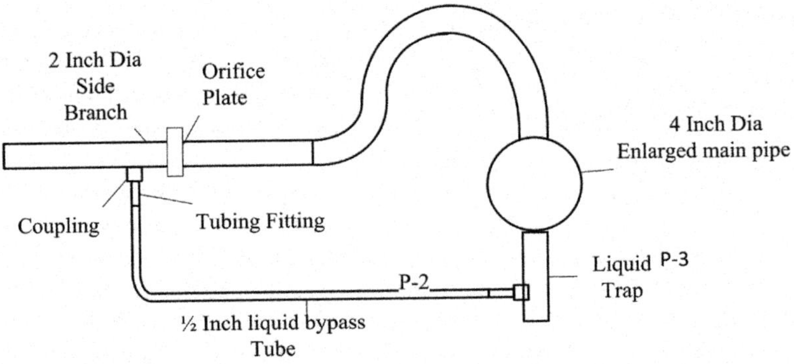

Fig. 3.15 SpliTigator T, Redrawn from: [35]

Fig. 3.16 Phase separation in two types of T-junctions, Redrawn from: [35]

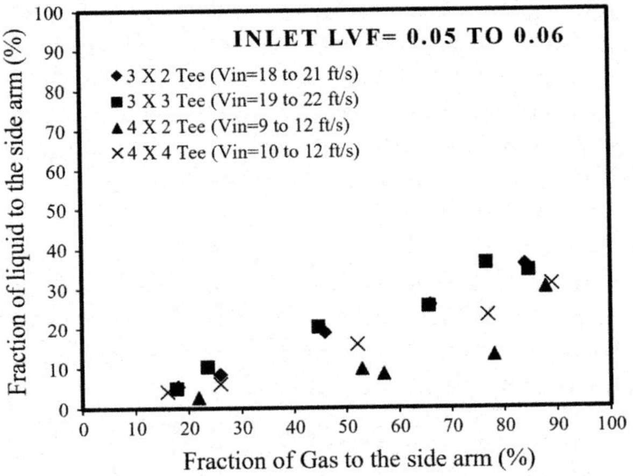

Fig. 3.17 Phase separation in regular, reduced, and spliTigator, Redrawn from: [35]

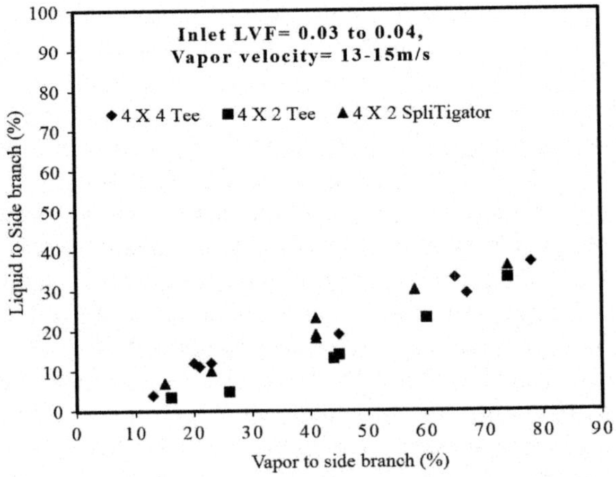

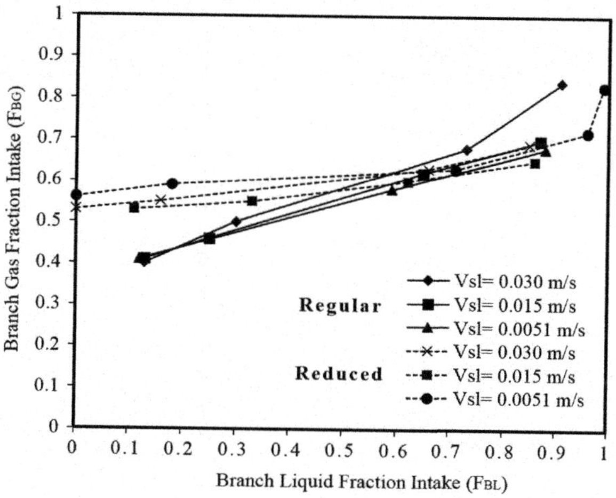

Fig. 3.18 Liquid and gas take off at 5° vertical side arm, Redrawn from: [11]

The experimental investigations conducted by Reimann et al. [36] examined the impact of branch angle and side arm diameter on separation performance. The researchers examined DR of 1, 0.52, and 0.2 to investigate the characteristics of stratified, slug, and annular flow regimes. The researchers made an observation that a decrease in diameter ratios led to an increase in phase separation performance or phase separation efficiency. The results provided by the author is consistent with the findings of other researchers in the existing body of literature. It suggests that a decrease in the diameter ratio leads to a reduction in the size of the side arm cross-sectional area, known as "zone of influence." Consequently, the decrease in size is associated with a decrease in the amount of liquid take-offs. The researchers developed the notion that as the ratio of diameters decreases, there is a corresponding increase in the efficiency of phase separation.

3.6.3 *Breakthrough by Walter et al. [37]*

The experimental investigations conducted by Buel et al. [38] focused on analyzing the pressure drop right at the junction. In their study, Walter et al. [39] furthered the existing research by examining the impact of diameter ratio on phase separation within a T-junction. Specifically, they investigated two different diameter ratios, namely 0.2 and 0.5. The authors conducted a comparative analysis of their findings in relation to the studies conducted by Buel et al. [38], Hart et al. [40], and Azzopardi [41]. The researchers' results presented a surprising contradiction to the prevailing understanding of the impact of diameter ratio. According to the data presented in Fig. 3.19, it can be observed that there is a notable decrease in the liquid take-off to the side arm as the diameter ratio decreases from 1 to 0.5. This finding is consistent with the established concept in this field. A comparable pattern was noted in the

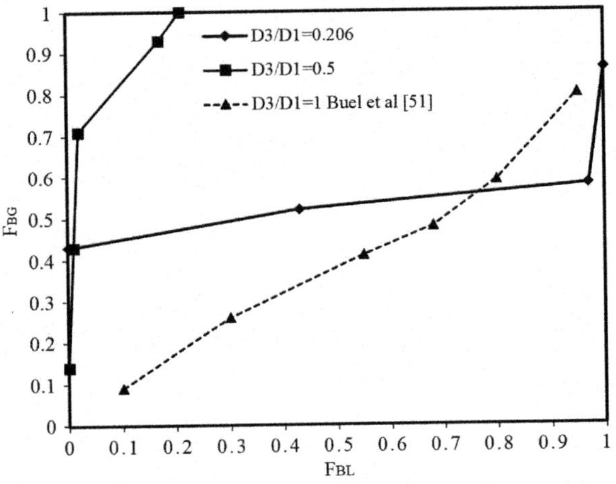

Fig. 3.19 Separation performance for different ratios of the diameters: Gas inlet velocity = 2.7 m/s, Liquid inlet velocity = 0.0021 m/s, Redrawn from: [37]

case of diameter ratios ranging from 0.5 to 0.2, albeit exclusively for lower rates of extraction. Nevertheless, with the escalation of the extraction ratio, there was a noticeable rise in the liquid take off, which contradicted the anticipated inverse relationship between the DR and separation performance. Although the researchers acknowledged the presence of uncertainty and disagreement in their observations, they did not specify the precise numerical value at which the inversely proportional relationship between the DR and the reversal of the trend occurred.

In their study, Walters et al. [39] examined the alterations in pressure drop pertaining to both the run arm and side arm. The data presented in their study indicate that there is a strong correlation between the pressure drop and the superficial velocities and mass split ratio of the respective phases. The researchers discovered that altering the diameter ratio from 1.0 to 0.206 did not yield a substantial decrease in pressure within the main arm. However, it did have a significant impact on the pressure drop within the side arm. Modifying the diameter ratio and mass split ratio induces changes in the pressure field within the junction. Additional variables, such as phase superficial velocities and mass split ratio, also play a role in influencing alterations in pressure drop within the side arm. It can be inferred that the pressure field exhibits a strong reliance on the superficial velocities of the phases and the mass split ratio.

In their study, Azzopardi [42] conducted an analysis to examine the impact of diameter ratio on phase split in regular and reduced T-junctions. The diameter ratios considered were 1, 0.67, and 0.33, while the main pipe diameter used for the air–water flow was 0.038 m.

The data presented in Figs. 3.20 and 3.21 clearly indicate that altering the diameter ratio from 1 to 0.67 led to a notable and statistically significant variation in phase separation across all extraction ratios. Nevertheless, the observed disparity in phase separation between a diameter ratio of 0.67 and 0.33 did not exhibit a noteworthy level of significance when compared to the disparity between 0.67 and 1. The data clearly

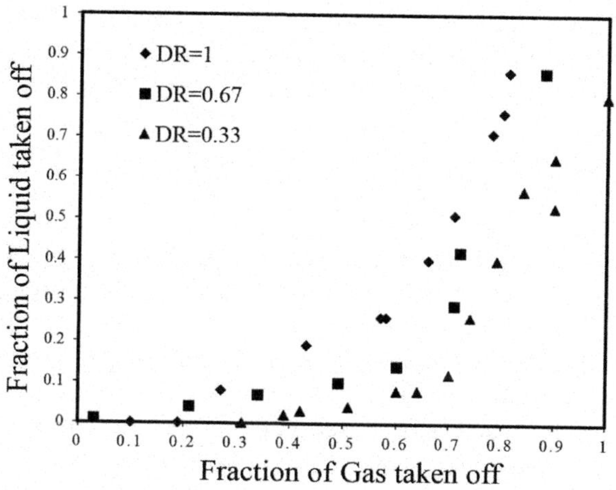

Fig. 3.20 Effect of diameter ratio for stratified flow, Redrawn from: [42]

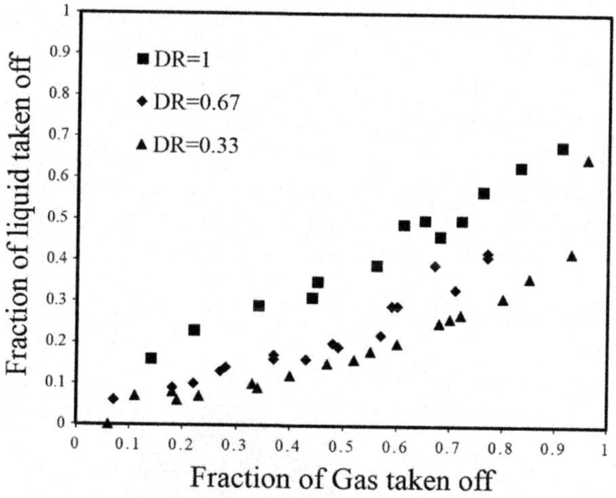

Fig. 3.21 Effect of diameter ratio for annular flow, Redrawn from: [42]

indicate that for lower extraction ratios, a diameter ratio of 0.33 continues to exhibit an inverse proportionality trend. However, as the extraction ratios increase, there is a notable rapid increase in liquid take-off. Azzopardi [42] conducted a comparative analysis of his findings in relation to the works of other researchers [19, 37, 38].

3.6.4 Agreement with Walter et al. [37] Finding

Pao et al. [43] conducted numerical simulations on T-junctions with diameter ratios of 1, 0.6, and 0.167. The smaller diameter ratio of 0.167 showed high liquid take off as compared to high DR. The reason is the diversion of gas into the side arm at higher extraction ratios, creating a pressure drop that leads to the liquid take off. The observation aligns with the findings of Walter et al. [37] regarding higher liquid take off at high mass split ratios. Saieed et al. [25] proposed a new key performance index to assess the performance of the T-junction.

Saieed [44] experimentally investigated the separation performance using five different diameter ratios of 1, 0.67, 0.52, 0.27, and 0.2. Obtained results showed a significant drop in phase liquid carryover (PLC) for stratified wavy flow as the diameter ratio decreased, as seen in Figs. 3.22, 3.23, and 3.24. The diameter ratio of 0.67 was found to be the most favorable, exhibiting high liquid carryover and relatively lower phase liquid carryover at all extraction ratios. Similar results were reported for slug flow, indicating optimum diameter ratios from 1 to 0.67 for high separation efficiency. The study by Saieed et al. [27] extended these experimental results, which were consistent with the numerical simulation findings of Pao et al. [45].

Tengku-Bazli and Pao [46] also reported similar results, concluding that decreasing the side arm diameter ratio increases phase liquid carryover. They used diameter ratios of 1, 0.67, 0.52, and 0.27 and suggested that it is possible to reduce liquid take off by increasing liquid phase velocities or by decreasing gas velocities.

Saieed [44] examined the pressure difference across the junction and found that it significantly increased as the DR decreased. This is attributed to the decrease in overall volume for fluid flow, resulting in higher velocities and, consequently, higher pressure drops. The pressure drop was observed to increase when transitioning from stratified wavy flow to slug flow, indicating its dependence on phase velocities and mass split ratios, consistent with the results presented by Walter et al. [39] and Casey [37].

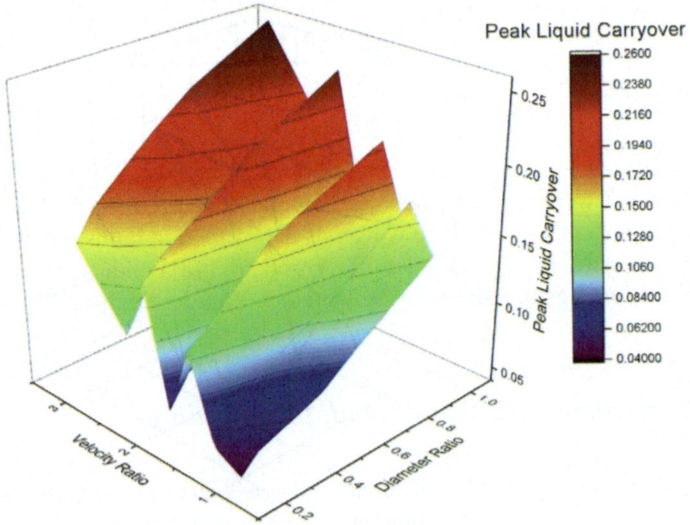

Fig. 3.22 Influence of DR on Peak Liquid Carryover [44]

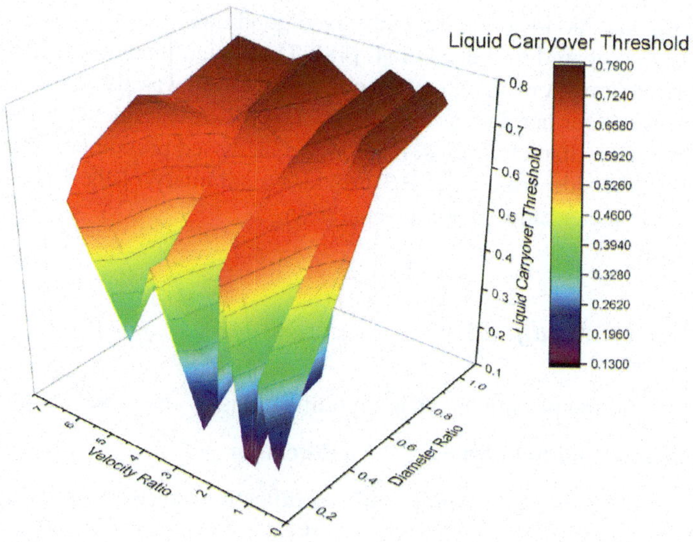

Fig. 3.23 Relationship between diameter ratio and LCT [44]

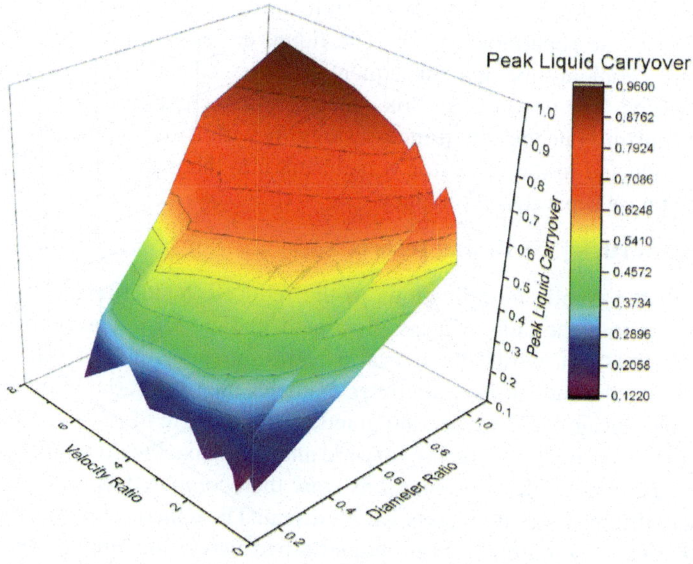

Fig. 3.24 Relationship between diameter ratio and PLC [44]

Abd-Hadi and Pao [47] obtained improved good outcomes for a T-junction with a DR of 0.27 compared to Saieed et al. [48]. The design included a side arm and a separate branch arm. A reverse trend in phase separation was also observed when as the diameter ratio further decreased from 0.27 to 0.2. It was due to the acceleration of air with smaller diameter ratios, resulting in increased pressure drop and pulling of liquid into the side arm. Table 3.3 provides a summary of the literature on the influence of diameter ratio on phase separation.

3.6.5 Design Modifications

Followings were the prominent design modifications by renowned scholars.

(a) **Katsaounis and Schultheiss** [7] **Modification**

Katsaounis and Schultheiss [7] proposed a modification aimed at improving the phase separation capabilities of T-junctions. A vertical conical tube was introduced at the aperture of the side arm. The conical tube exhibited an increase in diameter, ranging from 52 to 82 mm. In the horizontal T-junction with a vertical side arm, the main arm diameter measured 0.203 m. The aforementioned alteration was referred to as the dynamic phase separator system. The liquid extracted from the side arm was collected in a conical tube and subsequently redirected into the main arm through a backflow pipe. The objective of this concept was to enhance the process of phase separation and mitigate the occurrence of slugging within the pipeline. Reintroducing the liquid collected in the separator vessel back into the main arm has the potential to enhance phase separation efficiency.

(b) **Butterworth's** [55] **Tubing Inserts**

Butterworth [55] made modifications to the reduced T-junction by incorporating tubing inserts into the side arm, as illustrated in Fig. 3.25. The main arm had a diameter of 0.038 m, while the side arm had a diameter of 0.025 m. The tubing inserts possessed dimensions of an outer diameter measuring 0.025 m and an inner diameter measuring 0.022 m, accompanied by a cut angle of 45°. The inserts were arranged in two orientations: facing forward and facing backward relative to the main flow. The outcomes of these orientations were then compared to each other, as well as to the condition where no inserts were present. The experiments were carried out in three distinct flow regimes: stratified, stratified-wavy, and annular. Nevertheless, the implementation of tubing inserts did not yield a substantial enhancement in phase separation efficacy. In fact, in certain instances, the presence of tubing inserts resulted in a reduction of the overall efficiency of phase separation. The implementation of inserts oriented in a backward-facing manner resulted in enhanced phase separation within annular and stratified flow regimes, whereas inserts oriented in a forward-facing manner exhibited characteristics more akin to an even phase flow distributor.

Table 3.3 Summary of literature (effect of diameter ratio on phase separation)

Authors	DR	Working fluids	Superficial gas velocities (m/s)	Superficial liquid velocities (m/s)	Flow regime
Azzopardi and Whalley [28]	0.2, 0.4, 0.6	Air–Water	4.4–43.5	0.016–0.11	Ann, Bub, Ch
Azzopardi and Freeman [29]	0.8, 1	Air- Water	4.4–43.5	0.016–0.11	Ann
Shoham et al. [31]	1, 0.5	Air- Water	0.0029–0.059	6.1–26	St, Ann
Azzopardi and Smith [13]	1, 0.67, 0.33	Air–Water	0.75–25	0.008–0.067	Ann, St W, Sl
Mudde et al. [34]	0.43	Air–Water	0.05–0.15	0.5–1.5	St, St W, Bub
Marti and Shoham [11]	0.5	Air–Water	6.1	0.0051–0.059	St W
Griston et al. [35]	1, 0.67, 0.5	Steam-water	–	–	St, Ann
Reimann et al. [49]	1, 0.52, 0.2, 0.08	Air–Water	1–40	0.1–4	Sl, Bub, Ann
Walters et al. [37]	0.5, 0.2	Air–Water	2.6–40.7	0.0021–0.1820	St, St W, Ann
Azzopardi [42]	0.8, 1	Air–Water	4.4–43.5	0.016–0.11	Ann
Wren and Azzopardi [50]	1, 0.6	Air–Water	12–24	0.0535–0.31	St W, Ann
Tae and Cho [51]	1, 0.72, 0.44	R22 liquid–gas	–	–	Ann
Pao et al. [43]	1, 0.67, 0.167	Air–Water	0.816–2.45	0.03	St
Saieed et al. [25]	1.0, 0.52, 0.67, 0.27, 0.20	Air–Water	0.178–0.288	0.094–0.176	St W
Saieed et al. [27]	1, 0.67, 0.52, 0.27, 0.2	Air–Water	0.648–1.4	0.28–0.84	Sl
Tran et al. [5]	1 to 0.6	Air–Water	0.186–0.558	4–8	Sl
Memon et al. [52]	1, 0.67, 0.5, 0.27	Air- Water	0.20–1.16	0.40–0.53	Sl
Memon et al. [53]	01/ 0.52, 01/0.20 0.67/0.52, 0.67/0.27 0.5/ 0.27	Air–Water	0.388–0.62	0.252–0.582	St W, Sl

(continued)

Table 3.3 (continued)

Authors	DR	Working fluids	Superficial gas velocities (m/s)	Superficial liquid velocities (m/s)	Flow regime
Pao. and Memon [4]	$DR_1 =$ 1,0.67 $DR_2 = 0.2$, 0.27, 0.4, 0.5, 0.67, 0.8, 1	Air–Water	0.648–1.195	0.28–0.84	Sl
Azad and Pao [54]	$DR_1 = 0.55$ $DR_2 = 0.32$	Air–Water	0.8 and 1.2	1.5 and 2.7	Sl
T. Bazli and Pao [46]	1, 0.6, 0.5, 0.27 and 0.2	Air–Water	0.12 and 1.12	0.4 and 0.53	Sl

Whereas: DR = Diameter Ratio, St = Stratified, St W = Stratified Wavy, Sl = Slug, Bub = Bubbly, Ch = Churn

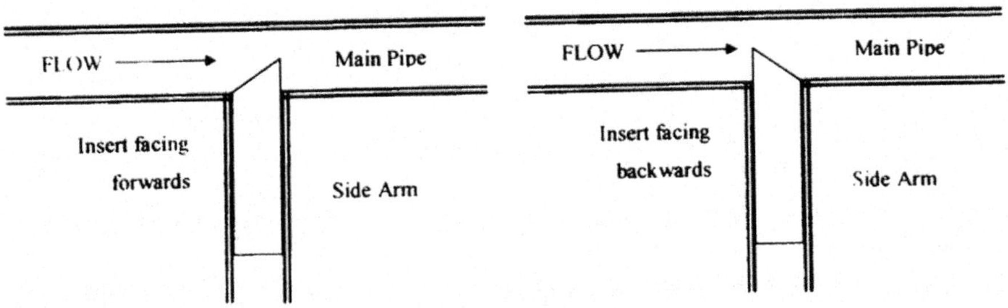

Fig. 3.25 Schematic diagram of the inserts used by Butterworth [55]

(c) **Baffles, Weirs, and U Bends**

The research conducted by Fouda and Rhodes [56], Azzopardi and Smith [57], and Wren [6] examined the impact of different alterations on the phase separation dynamics of T-junctions, particularly in relation to annular and stratified flow patterns. The implemented alterations encompassed the incorporation of baffles, weirs, and U-bends in various orientations downstream of the modified T-junction in order to augment the separation efficiency of phases.

Fouda and Rhodes [56]:

- Baffles were strategically positioned at the lower section of the T-junction, utilizing three distinct lengths: 1/4D, 1/2D, and 3/4D, with D representing the diameter of the main arm, as mentioned in Fig. 3.26.
- The objective was to attain equitable phase distribution between the two outlets, specifically the main pipe and the vertical upward side arm.
- The experiments were conducted within the annular flow regime.

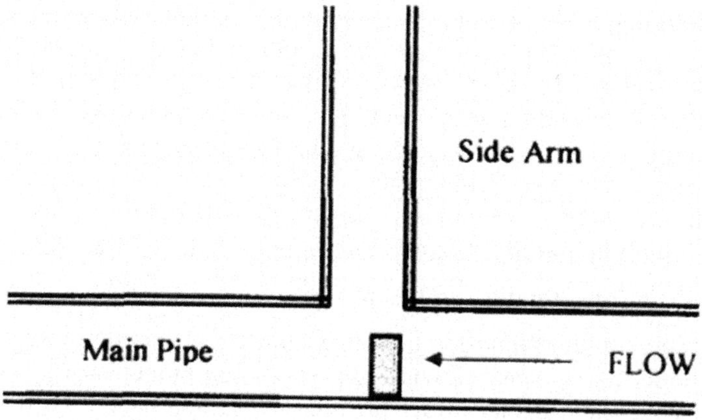

Fig. 3.26 Baffle Position in T-junction design of Fouda and Rhodes [56]

- The researchers discovered that the baffles with a diameter of 1/4D and 1/2D did not yield any statistically significant impact on the distribution of phase within the T-junction.
- Nevertheless, the 3D baffle did have an impact on the distribution of phase to a certain degree.

Azzopardi and Smith [57]:

- The researchers expanded upon the study by examining the impact of weir/baffles and a 90° bend downstream of the modified T-junction.
- The main arm maintained a horizontal orientation, while the orientation of the side arm transitioned from horizontal to vertical.
- The experiments were conducted to investigate annular or semi-annular flow regimes.
- The researchers observed that the presence of weir/baffles did not yield any discernible impact on the phase separation dynamics of the T-junction in both annular and semi-annular flow configurations.
- Nevertheless, a minor impact was noted in the stratified flow regime.

Wren [6]:

- The present study examines the impact of a U-bend configuration positioned downstream of both the run arm and side arm in a vertically downward reduced T-junction.
- The presence of a U-bend on the downstream side resulted in enhanced phase separation, resulting in an increased extraction of liquid into the vertically downward side arm in both stratified and annular flow regimes.
- The prevention of gas phase infiltration into the side arm was facilitated by the presence of the liquid level in the U-bend, thereby enhancing the efficiency of phase separation.

The results reported by Azzopardi and Smith [58] presented an intriguing contrast to the findings of Wren [59]. Specifically, Wren [59] proposed that the observed effect could be attributed to a change in the orientation of the side arm, from horizontal to vertical. The observed discrepancy underscores the intricate nature of phase separation phenomena in T-junctions and the multifaceted impact that various factors can exert on the outcomes. Furthermore, it underscores the significance of comprehending the precise flow conditions and geometric configurations in order to attain the most favorable phase separation in T-junctions across various applications.

(d) **Wren's** [6] **Investigation on Inserts**

In a subsequent study, Wren [6] conducted further research on the application of inserts in order to augment the phase separation capabilities of T-junctions. The present study examined the impact of insert protrusion depth, insert cut angle, and insert orientation on regular and reduced T-junctions. The experimental results demonstrated that the utilization of backward-facing inserts led to enhanced phase separation in both annular and stratified flow regimes. Conversely, the implementation of forward-facing inserts resulted in reduced phase separation and exhibited characteristics more akin to an even phase flow distributor. The experimental outcomes were consistent for insert cut angles of both 30° and 45°, as the observed profile of the incoming two-phase flow remained unaltered irrespective of the specific angle.

(e) **Reimann's** [60] **Gas Separator Designs**

Reimann [60] introduced novel gas separator designs for stratified and slug flow, which are positioned at the side arm of the T-junction, as depicted in Fig. 3.27. The design of the separator aimed to ensure that its volume did not exceed 10 times the volume of the branch arm. The liquid that was extracted into the side arm was subsequently collected in the separator vessel. In design A, the liquid was then directed back into the main horizontal pipe through the same branch. Design B involved the external occurrence of liquid backflow through an aperture located in close proximity to the lower section of the horizontal pipe. This design capitalized on the static water head as a propelling force for the liquid to flow in reverse within the horizontal pipe. Nevertheless, the existing designs were unable to achieve complete phase separation, prompting the suggestion to employ a multistage separator system in order to achieve full phase separation.

(f) **Azzopardi et al.** [8] **Distillation Column Modification**:

A design modification was proposed by Azzopardi et al. [8] to address the issue of declining efficiency in a distillation column at an operational plant in the UK (Fig. 3.28). This decline was attributed to the continuous liquid carryover with the vapor phase, specifically at the T-junctions. The alteration entailed the adjustment of the downstream geometry of the side arm through the implementation of a U-bend. The U-bend functioned as a physical obstruction that impeded the movement of the gas phase, thereby mitigating the occurrence of liquid entrainment within the gas phase. The two streams, which contained a high concentration of components in

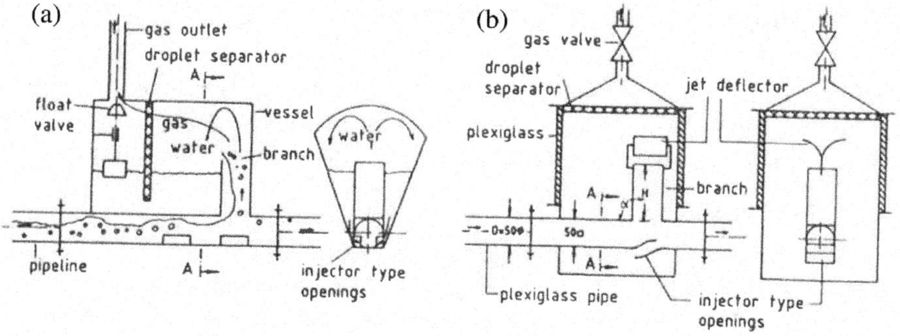

Fig. 3.27 Conceptional design new gas separator proposed by Reimann [60]

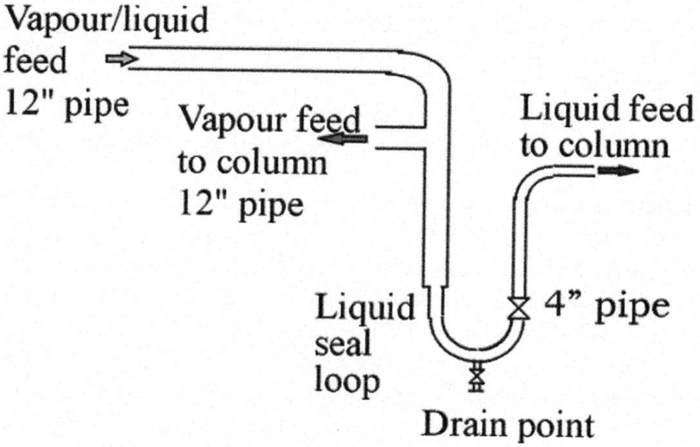

Fig. 3.28 Proposed design of T-junction partial phase separator [8]

each phase, were introduced into the column at precise locations in order to optimize the effectiveness of the separation process.

In general, these modifications were implemented to rectify particular flow characteristics and operational challenges encountered in diverse applications. Nevertheless, the efficacy of these alterations may differ based on the particular flow regime and operational circumstances, necessitating additional research and experimentation to enhance their performance.

3.6.6 Multiple Junction Combination

Multiple T-junction layouts and arrangements have been developed as a result of the requirement to increase the efficiency of phase separation in T-junctions. Investigations and configurations that stand out include some of the following:

(a) **Comb Separators**

Bevilacqua et al. [61] were the first to investigate the utilization of T-junctions as phase separators and enhance separation efficiency through the combination of different T-junctions in diverse arrangements and configurations. The concept of a "comb separator" was introduced, which entailed the integration of basic T-junctions and Y-junctions with vertically oriented upward side arms. The experiment was carried out under a pressure of 0.4 MPa, while the flow rate of the incoming liquid was adjusted within the range of 10–30 m3/hr. A series of two-phase separation experiments were conducted in order to optimize the configuration of the comb separator that yields the best results.

Figure 3.29 illustrates various arrangements of comb separators, encompassing both T&TY combinations (Fig. 3.9a) and combinations of Ts (Fig. 3.29b).

- The highest separation efficiency was attained by utilizing the T and TY configuration in Fig. 3.29a.
- The performance of the comb separators exhibited a decrease as the superficial velocity of the liquid was increased. This finding indicates that increased liquid velocities impede the phase separation capabilities of the comb separator.
- Full phase separation was achieved by reducing the downstream pressure and flow rates. Nevertheless, with the augmentation of flow rates, the efficacy of the comb separator exhibited a decline, thereby suggesting inherent constraints in its ability to manage higher flow rates.

The research conducted by Bevilacqua et al. [61] yielded significant findings regarding the potential enhancement of phase separation efficiency through the integration of various T-junction configurations. Nevertheless, it is crucial to take into account the flow rates and downstream pressure conditions in order to maximize the efficiency of these comb separators. Additional investigation and experimentation may be necessary to examine the viability of these configurations for particular applications and operational circumstances.

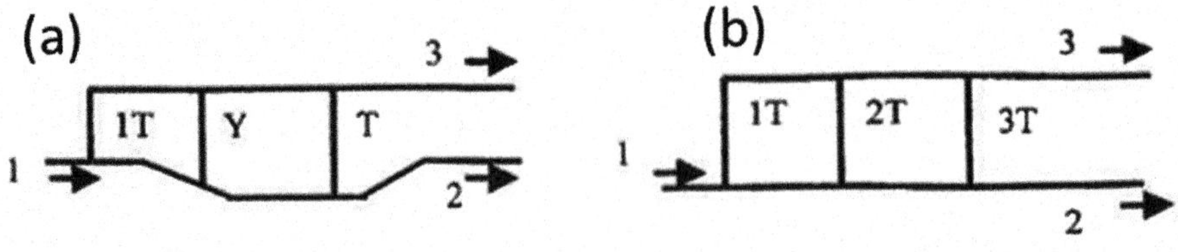

Fig. 3.29 Comb separators arrangement. **a** combination of T&TY; **b** combination of Ts [61]

(b) Two T-Junction System

In Wren's study [6], the objective was to establish distinct separation criteria within a T-junction system. The desired outcome was to ensure that the gas phase contained less than 10% of liquid, while the liquid phase contained less than 10% of gas. In order to accomplish this, Wren employed a sequential arrangement of two T-junction configurations. The initial T-junction featured a side arm oriented at a vertical upward angle of +90°, resulting in the generation of a stream abundant in gas. Conversely, the subsequent T-junction had a side arm oriented at a vertical downward angle of −90°, leading to the production of a stream predominantly composed of liquid. Figure 3.30 displays the schematic diagram of the two T-junction system.

- The separation criteria of having less than 10% liquid in the gas phase and less than 10% gas in the liquid phase were successfully met by Wren [59] in the combined T-junction system.
- Significant phase separation was achieved under conditions of low gas and liquid flow rates, indicating a high level of efficiency in separating the phases at these flow rates.
- Nevertheless, it was observed that with the increase in flow rates, the separation efficiency of the T-junction system exhibited a decline, ultimately reaching 85%, which fell short of the predetermined criteria. This implies that the system's ability to maintain the desired phase separation efficiency may be compromised when handling higher flow rates.
- In the context of annular flow, comparable findings were obtained, suggesting that the integrated T-junction system was successful in achieving phase separation in both stratified and annular flow regimes.
- The enhancement of phase separation efficiency was observed through the reduction in the diameter of the side arm in the second T-junction. This modification specifically facilitated the extraction of gas in the side arm, while maintaining a

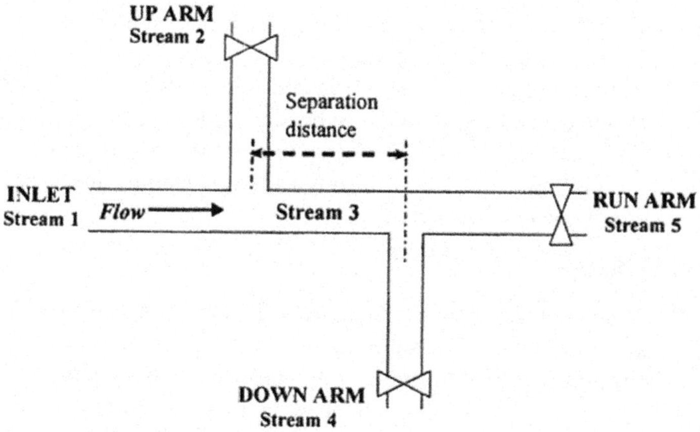

Fig. 3.30 Schematic diagram of the two T-junction system used by Wren [6]

consistent liquid take-off rate. This implies that modifying the diameter of the side arm has the potential to impact the behavior of phase separation, thereby offering an opportunity to enhance the efficiency of separation through optimization.

- In addition to the aforementioned research, Wren (22) conducted an investigation on the impact of separation distance between the two T-junctions, specifically at distances of 4D and 12D. However, the findings indicated that the separation distance did not exert a substantial influence on either the annular or stratified flow regimes. This observation suggests that the proximity of the T-junctions may not exert a significant influence on the phenomenon of phase separation in the specific system under investigation.
- In the study conducted by Wren [59], it was shown that the combination of T-junction configurations has the potential to effectively achieve specific phase separation criteria. Nevertheless, it is crucial to acknowledge the constraints of the system at elevated flow rates, prompting the need for additional investigation to enhance the system's performance under different operational circumstances. Furthermore, it is imperative to explore the impact of various factors, including geometry and fluid properties, in order to comprehensively comprehend and enhance the phase separation capabilities of the integrated T-junction system.

(c) **Gas–Liquid Phase Separator with STuVA Control**

STuVA-Controlled Gas–Liquid Phase separation: Priestman and Tippetts [62] suggested using a STuVA to regulate a gas–liquid phase separation (Fig. 3.31). A combination of liquid underflow and gas overflow, with STuVA maintaining an interface between the two phases, was how the separator controlled the flow of fluids. In order to keep the gas from entraining in the liquid-rich stream, this configuration helped maintain liquid impediment to the gas.

(d) **Multi-Tube T-Junction Separator**

The study conducted by Yang et al. [63] introduced a novel and innovative design called the "multi-tube T-junction separator," as seen in Fig. 3.32. This design utilized multiple T-junctions integrated into a single unit to achieve enhanced phase separation efficiency. The configuration consisted of a header connected to several T-junctions through connecting tubes or risers. Each T-junction acted as a partial phase separator, and as the gas–liquid mixture flowed through the connecting tubes, partial phase separation occurred at each T-junction.

Benefits and Findings:

- One of the main advantages of the multi-tube T-junction separator was its ability to achieve improved phase separation efficiency by utilizing multiple T-junctions within a single unit. The division and combination of gas and liquid phases through multiple T-junctions significantly enhanced the phase separation process.
- The researchers investigated various configurations with different numbers of connecting tubes. They found that the multi-tube T-junction separator achieved

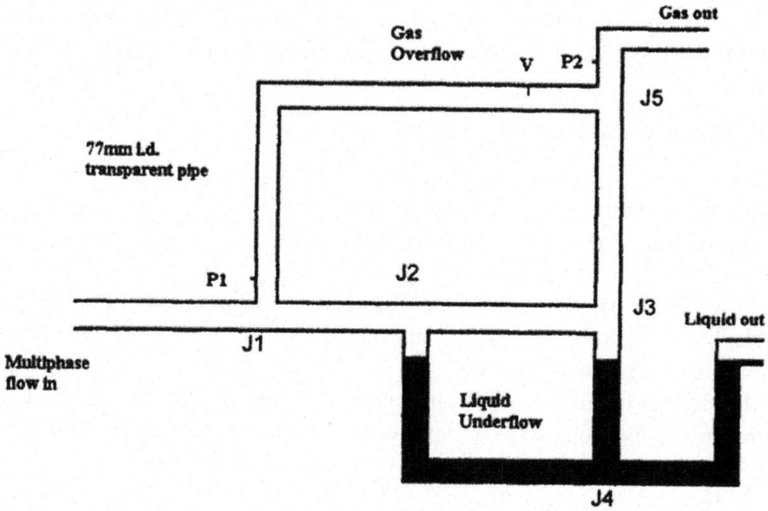

Fig. 3.31 Priestman and Tippetts [62] arrangement for T-junction compact separator

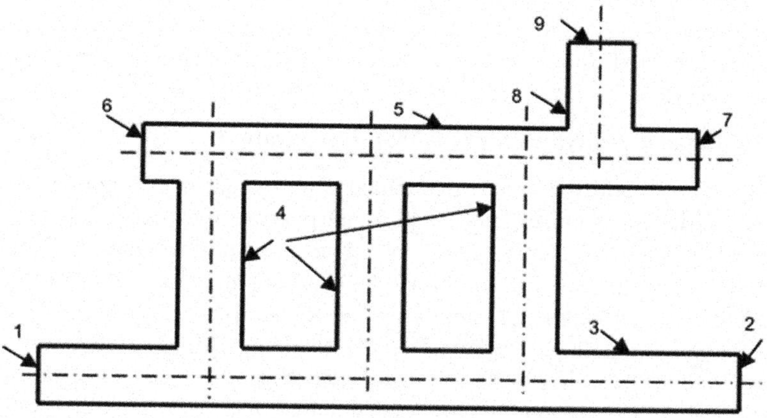

Fig. 3.32 Schematic diagram of multitube T-junction separator [64]

complete phase separation (gas and liquid phases that were nearly pure) in the case of stratified and plug flow when it contained five or more connecting tubes. This means that the output from the separator satisfied the separation criteria, demonstrating nearly perfect separation.

- The design's ability to achieve full phase separation makes it a potentially effective strategy for applications that require high separation efficiency, particularly in scenarios where near-perfect phase separation is necessary.
- The study highlighted that the performance of the multi-tube T-junction separator was influenced by the superficial velocities of both the gas and liquid phases.

At higher flow rates, the efficiency of the separator decreased. This indicates that operational constraints should be taken into account when implementing this design.

Implications and Future Prospects:

- The multi-tube T-junction separator shows promising prospects for various industrial processes that involve gas–liquid separation. Its ability to achieve near-perfect separation makes it suitable for applications where high separation efficiency is critical.
- However, the design's performance limitations at higher flow rates need to be carefully considered, and operational constraints should be taken into account during its implementation.
- Further research and optimization may be required to identify the most suitable operational conditions and configurations for specific applications, as well as to explore potential modifications or improvements to the design.

In conclusion, the multi-tube T-junction separator introduced by Yang et al. [63] presents an innovative approach to enhance phase separation efficiency. By combining multiple T-junctions within a single unit, the design achieves nearly perfect phase separation, making it a promising option for various industrial applications involving gas–liquid separation. However, operational constraints should be carefully considered, and further research is needed to optimize its performance for specific scenarios.

(e) **Combined T-Junctions for Oil–Water Separation**

Chen et al. [65] conducted a study that aimed to investigate the phenomenon of two-phase separation in combined T-junctions for the purpose of oil–water separation. The configuration employed in this study involved the integration of multiple T-junctions, collectively referred to as the combined T-junction configuration. The researchers focused their investigation on examining the impact of varying side arm separation distances within these interconnected T-junctions. The experimental setup was designed and implemented in order to investigate the research question. The configuration of the combined T-junction is depicted in Fig. 3.33. The researchers employed uniform spacing intervals for the T-junctions, with each junction represented by varying lengths, specifically 4D, 8D, 12D, 16D, 20D, 24D, and 28D.

In the experiments, the vertical height of all the side arms was established as 4D, where D denotes the diameter of the main pipe. For the purposes of the experiments, the value of D was fixed at 0.04m.

- The findings of the study indicate that the efficiency of phase separation was observed to be positively correlated with both the distance between side arms and the height of the side arm in the combined T-junction configuration. This implies that increased distances between the T-junctions resulted in enhanced phase separation between the immiscible oil and water phases.
- Furthermore, the researchers made observations that, when subjected to fixed inlet boundary conditions, the utilization of split flow in the combined T-junction

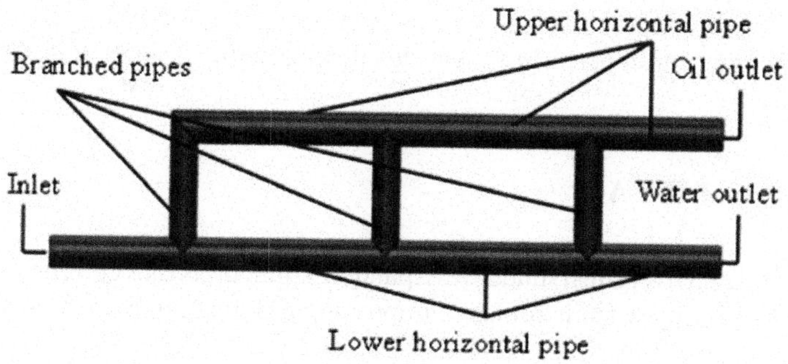

Fig. 3.33 Schematic diagram of combined T-junctions [65]

configuration effectively facilitated the process of phase separation between oil and water. This suggests that the structural arrangement of the combined T-junction configuration facilitated the effective segregation of phases.

- The implications and significance of the subject matter are of paramount importance.
- The results obtained from this research indicate that the optimization of both the side arm separation distance and height in a combined T-junction configuration can result in improved efficiency of phase separation in situations involving the separation of immiscible oil and water.
- The utilization of split-flow in the design of combined T-junctions has the potential to enhance the effectiveness of phase separation, thereby rendering it a highly promising strategy for a wide range of oil–water separation applications.

Potential Areas for Future Exploration:

- Future investigations may prioritize the examination of the most advantageous arrangements and operational parameters for the integrated T-junction configuration in order to attain enhanced efficiency in phase separation.
- Furthermore, it would be advantageous to explore the suitability and potential for expansion of this integrated T-junction arrangement in practical oil–water separation applications.

In summary, the research conducted by Chen et al. [65] regarding the utilization of combined T-junctions for the purpose of oil–water separation has indicated that the optimization of the separation distance and height of the side arm within the configuration can yield notable enhancements in the efficiency of phase separation. In addition, the incorporation of split flow within the combined T-junction design was instrumental in facilitating effective phase separation. The aforementioned findings provide significant insights that can contribute to the improvement of oil–water separation processes, as well as pave the way for additional research and potential applications in industrial settings. These studies show that attempts are being made to improve the phase separation properties of T-junctions by adopting different designs and arrangements to satisfy varied operational needs and separation criteria. Each proposed design has pros and cons, and more investigation and testing may be needed to improve how well it performs in various flow scenarios and applications.

3.7 State of the Art

Converging and Diverging (suddenly expanded T-junctions) are currently among the latest developments when it comes to improving the separation efficiencies.

3.7.1 Converging Design by Memon et al. [53]

Memon et al. [53] introduced a novel design methodology aimed at overcoming the constraints associated with T-junctions characterized by both higher and lower diameter ratios. A novel junction geometry was implemented, wherein the side arm was composed of two distinct diameter ratios, as illustrated in Fig. 3.34. These diameters are known as primary and secondary diameters. The main diameter ratios employed in the study were 1, 0.67, and 0.52, whereas, secondary diameter ratios observed in the slug flow were 0.52, 0.20, 0.52, and 0.27.

Separation efficiency of converging T-junction design was compared to that of a traditional T-junction by the researchers. The researchers discovered that the separation efficiency of the converging T-junction was 20%–35% greater as compared to regular T-junction.

One of the notable findings pertained to the impact of altering the main diameter while maintaining a constant secondary pipe diameter. The reduction of the main diameter from 1 to 0.5 resulted in a corresponding decrease in the liquid carryover (LCT) by approximately 7.23% and 14.1%, respectively. Furthermore, it was observed that the Peak Liquid Carryover (PLC) exhibited a 13% increase when the

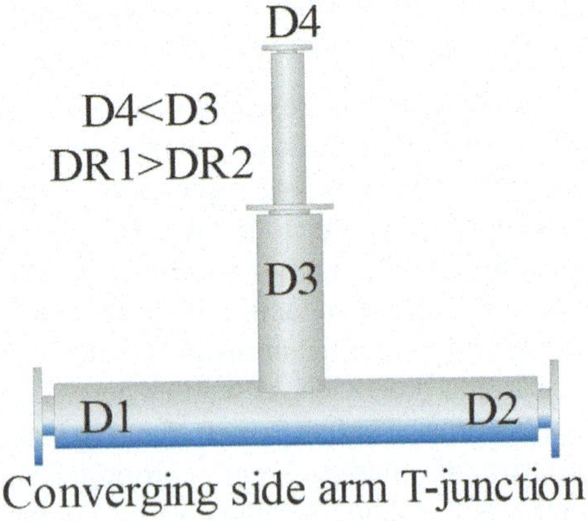

Fig. 3.34 Converging T-junction [53]

main diameter was reduced. The observed behavior can be ascribed to the reduction in diameter ratio, which consequently led to an increase in phase velocities. This increase in phase velocities resulted in a high pressure drop in the intersection, thereby resulting in liquid take offs.

The implementation of a reduced secondary diameter, while maintaining a constant main diameter, was found to be successful in improving the efficiency of phase separation. The graphical representation in Fig. 3.35 demonstrates the relationship between the secondary pipe diameter and the Peak Liquid Carryover (PLC). It is observed that a decrease in the secondary pipe diameter results in a significant decrease in the PLC. This decrease in PLC compensates for the disadvantage of having a larger diameter main pipe. The design modification yielded a significant 40% augmentation in the gas threshold and a 75% reduction in liquid take-off when compared to a standard T-junction.

The process of fallback of liquid from side arm as a result of momentum loss was further confirmed by Azad and Pao [54]. The study involved the implementation of numerical simulations, which revealed that during the initial 6-s period, the liquid take-off rate was observed to be greater in comparison to a standard T-junction configuration. The phenomenon described by Bernoulli's equation elucidates that when the diameter of the secondary pipe is reduced, there is an increase in the pressure drop. This results in the elevation of the liquid from the main arm and facilitates improved phase separation.

Pao and Memon [4] conducted a numerical analysis to evaluate the performance of the proposed converging T-junction design. The study employed Ansys Fluent software with a Volume of Fluid (VOF) model. The researchers examined primary

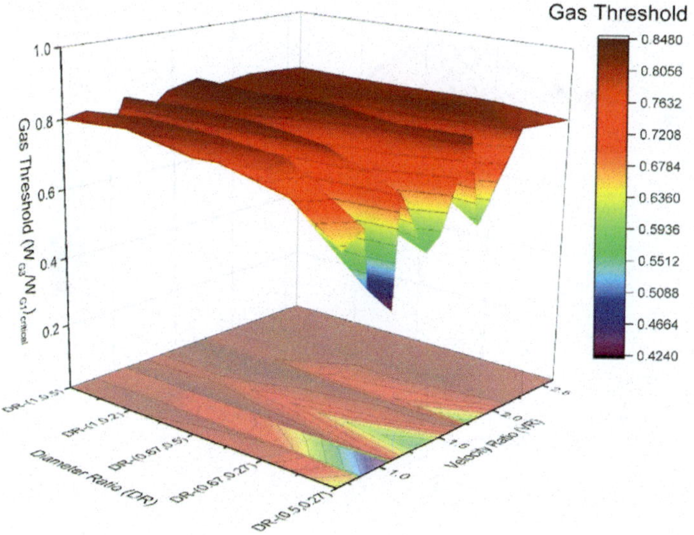

Fig. 3.35 Influence of DR on LCT [53]

DR of 1 and 0.67, while manipulating the secondary diameter ratios within the range of 0.2–1. The findings, as depicted in Fig. 3.36, demonstrated that the primary diameter side arm functioned as a typical T-junction, while the abrupt convergence of the secondary side arm impeded the flow of the liquid, resulting in its retraction toward the main arm. The researchers reached the conclusion that the most effective combinations of diameter ratios were 1/0.5 and 0.67/0.4, resulting in a phase separation efficiency exceeding 90%. It is important to acknowledge that the simulations conducted were restricted to the flow of air and water. Therefore, additional experimentation involving alternative working fluids is necessary.

Numerical simulations are of paramount importance in the investigation of various phenomena, as they offer significant insights while simultaneously reducing time and equipment expenses. However, it is imperative to corroborate simulation outcomes through empirical experimentation. Zhang et al. [66] reported comparable findings between simulations and experimental observations, albeit with minor deviations within a particular flow range. These discrepancies were attributed to equipment roughness resulting from low machining precision, a factor that was not considered in the simulation.

Various methodologies have been utilized in the design of simulation workbenches, including the implementation of perturbations within the pipe and the imposition of initial boundary conditions to confine the liquid and gas phases solely occupying half of the pipe. These methodologies facilitate the prediction of phase separation with improved precision and authenticity, thereby augmenting the dependability of simulation outcomes.

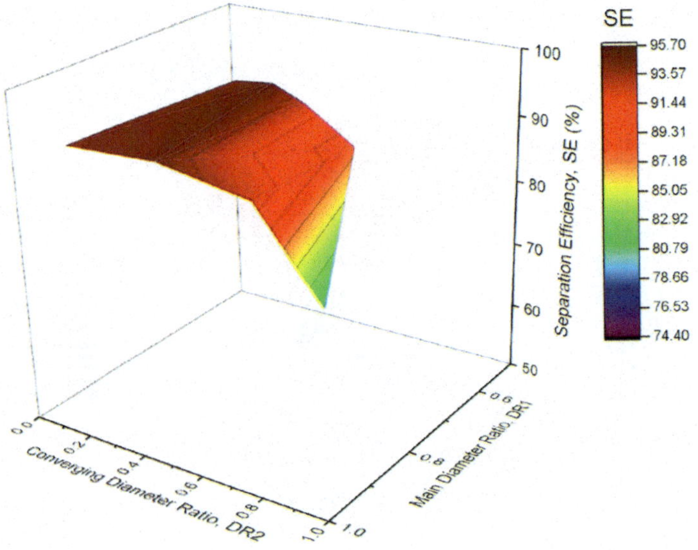

Fig. 3.36 Effect of diameter ratio on SE [4]

3.8 Diverging Design by Ejaz et al. [3]

In this numerical inquiry, the researchers directed their attention toward examining the efficacy of five distinct suddenly expanded T-junctions across a range of operational circumstances. In order to conduct a comprehensive evaluation of their behavior, each T-junction underwent testing for eight distinct combinations of velocity ratios, resulting in a total of 200 simulations.

The researchers conducted an analysis on the phase separation efficiency of the T-junctions in each simulation. This analysis involved the consideration of five data points, with each point representing a consistent mass split ratio. The mass split ratio pertains to the allocation of flow rate based on weight in the side arm. It was systematically altered at values of 0.2, 0.4, 0.5, 0.6, and 0.8 for each T-junction, while keeping the velocity ratio constant.

The velocity ratio is a significant factor in the process of gas–liquid separation, as it represents the relationship between the superficial velocities of the gas and liquid phases. The researchers conducted a study encompassing eight distinct velocity ratios in order to investigate a broad spectrum of flow conditions. This approach enabled them to examine the behavior of T-junctions across a range of flow rates. The comprehensive methodology employed in this study offers significant insights into the behavior of T-junctions in real-world situations characterized by potential fluctuations in flow rates.

In order to conduct the simulations, the researchers implemented measures to maintain an isothermal condition, thereby eliminating the influence of temperature fluctuations. This allowed for a concentrated examination of the phase separation phenomenon. Furthermore, the exit boundary condition was established to mimic atmospheric pressure, which is representative of common industrial scenarios where gases are released into the atmosphere following the separation process.

The impact of upstream and downstream diameter ratios (DR_1 and DR_2) on the performance of the T-junctions was also investigated by the researchers. The selected values for DR_1 were 0.2, 0.5, 0.8, 0.27, and 0.5, while the scrutinized values for DR_2 were 1 and 0.67, as given in Table 3.4. The researchers investigated the impact of different diameter ratios on the phase separation efficiency of T-junctions in order to examine the influence of geometrical configurations.

In addition, the researchers conducted an analysis of the behavior of T-junctions under various flow regimes. They specifically chose two distinct gas superficial velocities, namely 3.24 m/s and 5.975 m/s, and manipulated the liquid superficial velocities to include values of 1.4 m/s, 1.75 m/s, 2.45 m/s, and 3.491 m/s.

The study conducted a total of 200 simulations, which yielded a comprehensive and detailed comprehension of the response of suddenly expanded T-junctions under various flow conditions and geometrical configurations. The outcomes derived from these simulations will play a crucial role in enhancing the efficiency and cost-effectiveness of gas–liquid separation processes in diverse industrial contexts. Additionally, the knowledge acquired from this inquiry will function as a significant source of reference information for scholars and professionals engaged in the domain of

Table 3.4 Suddenly expanded T-junctions and velocity ratio combinations

T-Junction no	DR_1	DR_2	Full name
T_1	0.2	1	DR (0.2–1.0)
T_2	0.5	1	DR (0.5–1.0)
T_3	0.8	1	DR (0.8–1.0)
T_4	0.27	0.67	DR (0.27–0.67)
T_5	0.5	0.67	DR (0.5–0.67)
Velocity ratio no	V_G (m/s)	V_L (m/s)	VR value
VR_1	3.24	1.4	2.31429
VR_2	3.24	1.75	1.851
VR_3	3.24	2.45	1.32245
VR_4	3.24	3.491	0.9281
VR_5	5.975	1.4	4.26786
VR_6	5.975	1.75	3.4
VR_7	5.975	2.45	2.43878
VR_8	5.975	3.491	1.71154

gas–liquid separation and fluid dynamics. In its entirety, this quantitative investigation represents a noteworthy contribution to the progression of gas–liquid separation technologies and establishes the foundation for subsequent research and innovation in this domain.

Their research study demonstrated a strong resemblance between the geometry of the T-junction design and the flow loop facility employed in prior research conducted by Saieed [44]. In order to facilitate a comparative analysis, meticulously selected diameter ratios were utilized, which served as a direct contrast to the published data for converging T-junctions [4].

The geometric configuration comprised a mixer positioned at the far left, succeeded by the T-junction segment, as mentioned in Fig. 3.37. The mixer facilitates the combination of water and air, resulting in the formation of a mixture within designated sections measuring 0.25 m in length for both the air and water inlets.

The main arm diameter was of 0.074 m, which is equivalent to 3 inches. The lengths of the side arm pipes on the upstream and downstream were maintained at 0.5 m and 1 m, respectively, in accordance with prior research conducted on regular and converging T-junctions [4, 44]. A total of five distinct combinations of side arm diameter ratios were utilized while maintaining a consistent length for all arms. The diameter ratio combinations employed in the study were (0.2–1), (0.5–1), (0.8–1), (0.27–0.67), and (0.5–0.67), and these combinations were simulated in a three-dimensional model.

In order to enhance the precision of simulations and decrease the amount of computational resources required, the T-junction was divided into five distinct bodies while still being regarded as one part, as depicted in Fig. 3.37. The utilization of this modeling approach enabled the researchers to incorporate a hybrid meshing

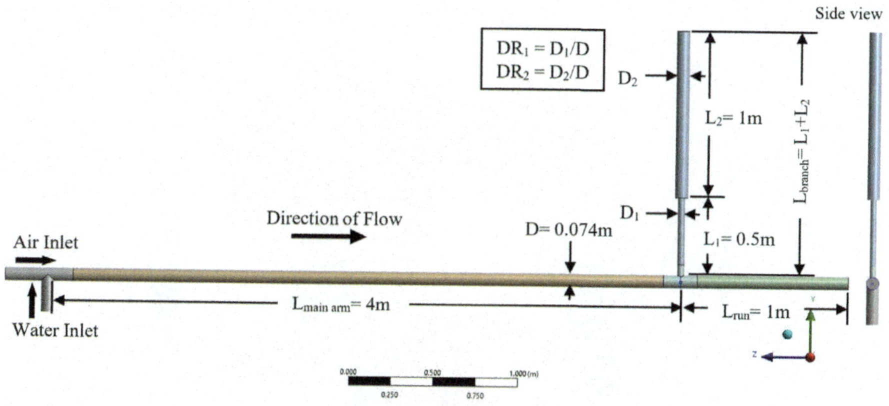

Fig. 3.37 T-junction design in ansys design modeler [3]

technique, thereby augmenting the precision of the solutions while simultaneously reducing the computational power required.

The objective of their study was to investigate the phase separation efficiency of five distinct T-junctions by employing 3D response surfaces. The surfaces presented in this study were derived from actual numerical data, thereby offering a comprehensive comprehension of the performance of each T-junction across a range of velocity ratios (VR_{1-8}).

After conducting an analysis of Fig. 3.38, it is apparent that T_3 exhibited the highest level of separation efficiency, reaching a peak value of 96.33% for VR_1. Furthermore, T_2 exhibited commendable performance, exhibiting a marginally lower level of efficiency in comparison to T_3, yet surpassing the remaining T-junctions. Nevertheless, the separation efficiency exhibited a decline across VR_{1-8} for T_{1-4}. In instances where the superficial velocity of air remained constant (VR_{1-4}), an increase in the superficial velocity of water led to a reduction in separation efficiency. This can be attributed to the characteristics of suddenly expanded T-junctions. In a similar vein, it was observed that maintaining a consistent air velocity while augmenting the water velocity in the VR_{5-8} range resulted in a decline in overall efficiency. This decline was particularly pronounced in the T_{1-4} subset, where the integrity of phase separation was compromised. Notably, both T_1 and T_4 failed to achieve any degree of phase separation under these conditions. T_5 also exhibited a decline in separation efficiency across VR_{1-8}, indicating certain discrepancies in its performance.

After conducting a comprehensive analysis of the suddenly expanded T-junctions, it became apparent that T_2 and T_3 consistently exhibited superior performance compared to the remaining T-junctions. These two junctions consistently achieved maximum separation efficiencies of 96.55% and 94.552%, respectively, across all velocity ratios (VR_{1-8}). Even when subjected to significantly high superficial velocities, both T_2 and T_3 demonstrated separation efficiencies exceeding 80%, thus highlighting their strong and reliable performance. Nevertheless, the separation efficiencies of T_1 and T_4 were not satisfactory across all velocity ratios, specifically for VR_{5-8},

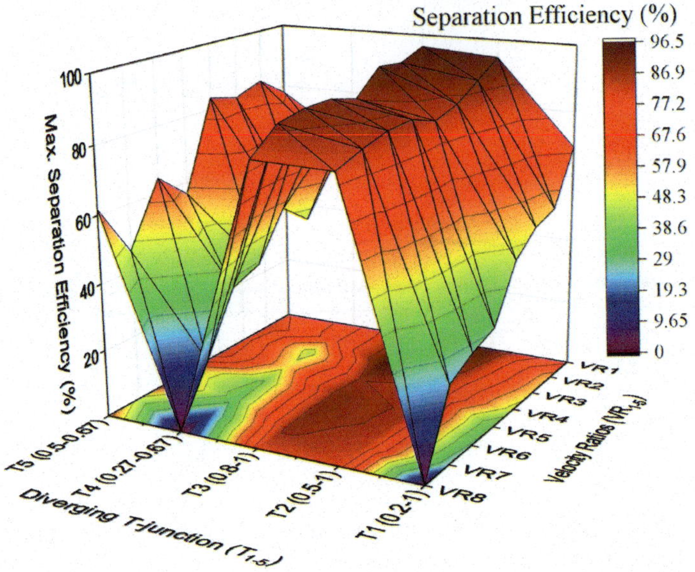

Fig. 3.38 Effect of diameter ratio combinations on separation efficiency (SE) [3]

where their separation efficiencies varied from 0% to 69.1%. These performance levels are deemed unsuitable for implementation in industrial settings.

As previously mentioned, the Liquid Carryover Threshold (LCT) denotes the upper limit of air content in the branch arm where no liquid is visibly present in the branch. Greater Liquid Carryover Threshold (LCT) values are advantageous in order to achieve improved phase separation. T-junctions T_2 and T_3 demonstrated elevated LCT values in comparison to the remaining T-junctions across all velocity ratios. This suggests that the utilization of T_2 (0.5–1) and T_3 (0.8–1) can effectively postpone the occurrence of liquid in the branch arm, rendering them more favorable options for industrial applications where the delay of liquid carryover is of utmost importance.

The researchers employed 3D response surfaces to acquire a comprehensive comprehension of the performance of five T-junctions in the process of phase separation. The response surfaces presented in their study were derived from empirical numerical data, enabling the researchers to evaluate the phase separation efficiency of each T-junction under various velocity ratios (VR_{1-8}). These response surfaces served as a visually informative tool for their assessment.

The response surface for the separation efficiency of the T-junctions is depicted in Fig. 3.39. Upon initial observation, it is evident that T3 exhibited the most notable separation efficiency, achieving a percentage of 96.33% at VR_1. T_2 also exhibited a commendable performance, demonstrating a slightly lower yet notable separation efficiency in comparison to the remaining three T-junctions. Nevertheless, upon thorough examination of the complete spectrum of velocity ratios, the researchers observed a decline in the separation efficiency of all T-junctions. In the case of

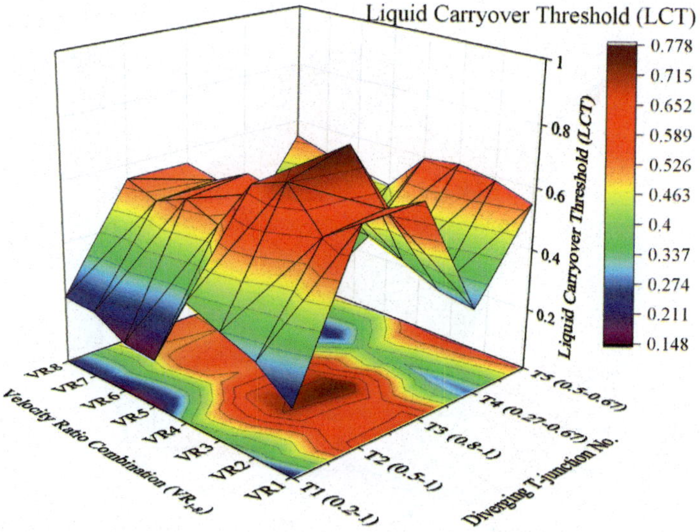

Fig. 3.39 Effect of diameter ratio combinations on Liquid Carryover Threshold (LCT) [3]

$VR_{1–4}$, with a constant air superficial velocity of 3.24 m/s, it was observed that altering the water superficial velocity resulted in a decrease in separation efficiency. In a similar vein, in the case of $VR_{5–8}$, where the air velocity remained constant at 5.975 m/s, an increase in water superficial velocity resulted in a further decrease in overall efficiency, primarily attributed to the impact of the elevated air superficial velocity. T_1 and T_4, specifically, demonstrated inadequate separation efficiency when subjected to higher velocity ratios ($VR_{5–8}$), thereby rendering them less appropriate for implementation in industrial settings.

The maximum Liquid Carryover Threshold (LCT) is a critical parameter in the phenomenon of phase separation, as it signifies the uppermost proportion of air within the branch arm before liquid becomes visibly present. The investigators employed a three-dimensional response surface, as depicted in Fig. 4.15, to examine the laminar coherent structures (LCT) in various T-junction configurations and velocity ratios. It is noteworthy that T_2 and T_3 consistently exhibited the highest LCT values among all T-junctions, irrespective of the velocity ratio. This implies that T_2 and T_3 exhibited a higher degree of efficacy in delaying the onset of liquid flow in the branch arm compared to the remaining T-junctions. In the case of $VR_{1–4}$ and $VR_{5–8}$, the longitudinal concentration gradient (LCT) exhibited an increase for all T-junctions, with the exception of T_2 and T_3, as the water superficial velocity had minimal impact on their LCT values. Nonetheless, a sudden decrease in local convective heat transfer (LCT) occurred in all T-junctions at VR_5, which can be attributed to the notably higher air superficial velocity of 5.975 m/s, in contrast to the 3.24 m/s observed at VR_4.

The PLC was also examined by the researchers utilizing a 3D response surface, as depicted in Fig. 3.40. Lower values of the phase-liquid concentration (PLC) signify a reduced proportion of liquid in the side arm, thereby rendering it advantageous

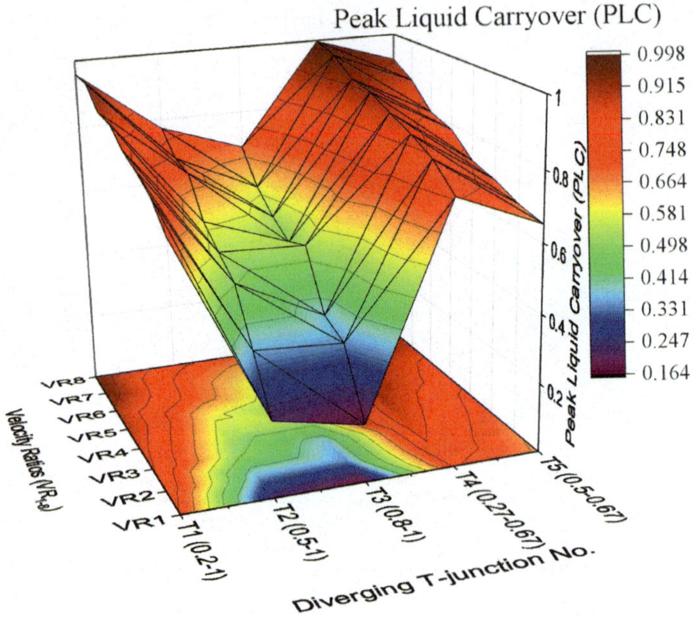

Fig. 3.40 Effect of diameter ratio combination on Peak Liquid Carryover (PLC) [3]

for achieving effective phase separation. T_2 and T_3 consistently exhibited lower Peak Liquid Carryover (PLC) values in comparison to the remaining T-junctions, regardless of the velocity ratios. The PLC values of the aforementioned choices did not surpass 0.7, rendering them advantageous options for industrial applications. In the case of VR_{1-4}, it was observed that the Peak Liquid Carryover (PLC) exhibited an increase across all T-junctions as the liquid superficial velocity was raised. The VR_{5-8} exhibited a comparable pattern, wherein the PLC demonstrated an increase as a result of the impact of air superficial velocities. The researchers reached the conclusion that in order to attain reduced PLC (Peak Liquid Carryover) values for a particular T-junction, it would be imperative to operate at lower air and water superficial velocities.

In brief, the utilization of 3D response surfaces has yielded significant insights into the efficacy of diverse T-junctions in the context of phase separation. This has facilitated a more comprehensive comprehension of T-junction performance, enabling the identification and selection of optimal T-junction configurations tailored to specific operational parameters and separation criteria within a range of industrial sectors. Furthermore, coded and uncoded correlations are given in the Appendix.

3.8.1 Comparison of the State of the Arts

Table 3.5 presents a thorough comparison of the numerical simulation and experimental findings obtained in their investigation (T_2 and T_3) with the published data of regular, reduced, and converging T-junctions. The experimental data utilized for the regular and reduced T-junction results were sourced from Saieed's publication [44]. The researcher conducted an investigation to examine the impact of diameter ratio on phase separation. For this purpose, pipes with a diameter of three inches were utilized for the main, run, and side arms. In contrast, the numerical simulation results of the converging T-junction were obtained from published data by Pao and Memon [4]. The researchers performed numerical simulations of slug flow utilizing the Volume of Fluid (VOF) model implemented in Ansys FLUENT.

By conducting a comparative analysis of the data presented in Table 3.5, they derived significant insights regarding the performance of T_2 and T_3 in relation to regular, reduced, and converging T-junctions, while considering similar velocity ratio conditions. The inclusion of this comparative analysis serves to substantiate the numerical simulations conducted in the current study and offers a more comprehensive understanding of the gas–liquid separation characteristics exhibited by T-junctions. Furthermore, it enables a thorough evaluation of the efficiency and effectiveness of T_2 and T_3 in industrial settings where the separation of gas and liquid phases is of utmost importance.

Figure 3.41 presents a detailed analysis of the separation efficiency for five different suddenly expanded T-junctions, in comparison to published data, across six distinct velocity ratios. The exclusion of VR_2 and VR_6 from this comparison was deliberate, as there is a lack of pertinent data available for these specific velocity ratios in relation to other T-junction types.

Table 3.5 Comparison of suddenly expanded T-junctions (T_2 and T_3) with published data of regular, reduced, and converging T-junction, in terms of Separation Efficiency (SE)

SE obtained from the published data (%)						SE achieved in present study (Suddenly Expanded T-junctions) (%)			
Inlet conditions		Saieed [44]		Pao and Memon [4] (Simulation)		Inlet conditions		T_2 (0.5–1)	T_3 (0.8–1)
V_a m/s	V_w m/s	DR (1)	DR (0.67)	DR (1–0.5)	DR (0.67–0.4)	V_a m/s	V_w m/s		
0.648	0.28	78	75	93	93	3.24	1.4	94.6	96.3
	0.49	78	76	94	92		2.45	85.2	94.4
	0.698	79	81	94	90.5		3.49	83.0	87.3
1.195	0.28	78	65	95	91	5.97	1.4	86.0	89.7
	0.49	77	70	93.5	92		2.45	83.6	88.0
	0.698	83	76	94	90.5		3.49	80.7	81.2

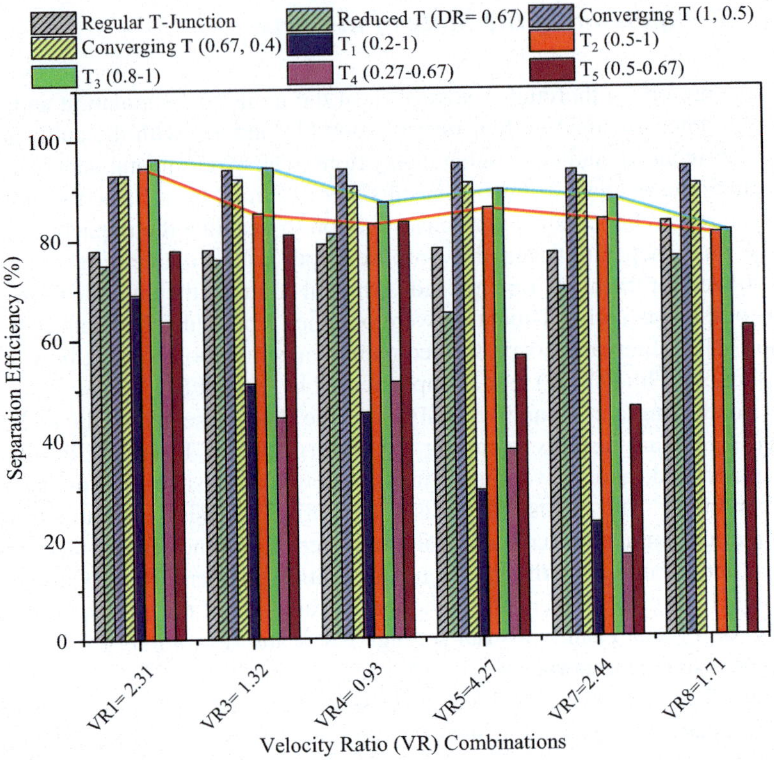

Fig. 3.41 Comparison of suddenly expanded T-junctions (T_{1-5}) with experimentation data of regular, reduced (Saieed [44]), and numerical simulation data of converging T-junction (Pao and Memon [4])

The T_1 and T_4 configurations consistently exhibited the lowest separation efficiency among all T-junction configurations, irrespective of the velocity ratio. Indeed, the separation efficiency exhibited an exceptionally minimal value of 0% at VR_8. The subpar performance can be ascribed to the notably low ratio of upstream diameters between T_1 and T_4. As a result, the re-entry of liquid into the side arm becomes exceedingly difficult. Moreover, the reduced diameter ratio contributes to increased superficial velocities and enhanced pressure drop in the branch arm, thereby further hindering effective phase separation.

In contrast, T_5 exhibited enhanced separation efficiency in comparison to both regular and reduced T-junctions across the initial three velocity ratios. Unfortunately, when considering different velocity ratios with higher air and liquid superficial velocities, the separation efficiency of T_5 experienced a significant decrease.

The phase separation data revealed that T_2 and T_3 consistently exhibited superior separation efficiency compared to other abruptly expanded T-junctions. This observation necessitates a comprehensive comparison with other types of T-junctions. In comparison to conventional and modified T-junctions, T_2 and T_3 have emerged as the most successful options, demonstrating improved separation efficiency due to their unique design incorporating a combined diameter ratio. This advantage was consistently maintained throughout all velocity ratios. Specifically, the regular and reduced T-junctions exhibited separation efficiencies that were lower than 83%. On the other hand, the implementation of abruptly enlarged T-junctions, such as T_3, yielded a significant enhancement of 13% in the efficiency of separation. Similarly, T_2 demonstrated a notable improvement of 11%.

When comparing converging T-junctions, it was observed that T_3 exhibited a notable maximum separation efficiency of 96.3%, while T_2 demonstrated a closely trailing efficiency of 94.6%. On the other hand, the T-junction with a convergence ratio of 1–0.5 achieved a peak separation efficiency of 95%. The separation efficiency of T_2 and T_3 was found to be higher than that of the converging T-junctions at the initial two velocity ratios. However, their efficiency declined as the velocity ratios increased. The observed difference can be ascribed to the presence of a converging section in the side arm of the T-junction, which functions as an impediment or constriction for the incoming fluid. As the liquid approaches the point of convergence, its momentum gradually decreases, leading to a regression in its motion. In contrast, the downstream pipe in the diverging T-junction does not impose similar limitations; instead, it extends the duration of liquid presence within the pipe. The phenomenon of liquid fallback is significantly constrained due to the lower upstream diameter ratio compared to the downstream counterpart.

The information derived from Table 3.5 is significant as it indicates that the phase superficial velocities observed in our study are considerably higher than those reported in existing literature. Before reaching definitive conclusions, it is imperative to engage in a comprehensive examination of this crucial aspect. Remarkably, it has been observed that even when subjected to high air and water flow rates, the diverging T-junctions (T_2 and T_3) consistently exhibit an impressive separation efficiency of over 80%. In direct opposition, the converging T-junctions exhibited marginally higher separation efficiency solely at extremely low superficial velocities. Therefore, it is crucial to investigate the efficiency of separation in converging T-junctions at elevated phase superficial velocities, which aligns with the parameters examined in our current research. Considering the frequent occurrence of elevated superficial velocities in industrial settings, both T_2 and T_3 present themselves as feasible options.

After comparing our findings with the simulated results of converging T-junctions, it is crucial to conduct a more comprehensive evaluation by comparing them to experimental data. The study conducted by Memon [67] primarily aimed to analyze the influence of the combined diameter ratio on phase separation in stratified and slug flow scenarios. The present study involved a comprehensive examination of published data pertaining to velocity ratios in order to facilitate a rigorous comparative analysis. It is worth mentioning that Memon's experimental configuration closely mirrored that of Saieed [44]. The main differentiation resided in the configuration of the side arm, characterized by a convergent design. Table 3.6 presents a comprehensive compilation of velocity ratios, encompassing data from our current investigation as well as previously published studies, thereby enabling a straightforward comparison. In particular, we have chosen three T-junction configurations that demonstrate higher separation efficiencies in comparison to their counterparts, for the purpose of conducting a detailed analysis.

In Table 3.6, thorough comparison is given among suddenly expanded T-junctions (T_2 and T_3) and previously published experimental data on converging T-junctions. The table presents a comparative analysis of SE values, providing valuable insights into the performance of various configurations under different inlet conditions.

The experimental data conducted by Memon [67], which includes a range of inlet conditions, are considered a benchmark in the field. The standard error (SE) values derived from the data are compared to the SE values obtained in our current study for T2 and T3, considering the same inlet conditions. The parameters encompassed by the inlet conditions consist of Va, denoting the superficial velocity of air, V_w, representing the superficial velocity of water, and various combinations of diameter ratios (DR).

The comparison elucidates captivating observations. It is worth noting that T_1 and T_4 demonstrate comparatively lower separation efficiencies in comparison to converging T-junctions. In contrast, T_5 exhibits a comparable level of separation efficiency to that of converging T-junctions when subjected to specific velocity ratios, specifically VR_1, VR_3, and VR_4. Nevertheless, when the velocity ratios increase toward higher superficial velocities, T_5 undergoes a substantial decrease in separation efficiency.

In order to further explore this comparison, Fig. 3.42 provides a more comprehensive depiction. The graph illustrates the disparity between the suddenly expanded T-junctions (T_{1-5}) and the experimental findings of converging T-junctions as reported by Memon [67]. The presented graphical representation provides a comprehensive depiction of the diverse separation efficiencies observed in various T-junction configurations and velocity ratios.

Table 3.6 Comparison of suddenly expanded T-junctions (T_2 and T_3) with published experimentation data of converging T-junction, in terms of Separation Efficiency (SE)

SE obtained from the published data (%)					SE achieved in present study (suddenly expanded T-junctions) (%)			
Inlet conditions		Memon [67] (experimentation results)			Inlet conditions		T_2 (0.5–1)	T_3 (0.8–1)
V_a m/s	V_w m/s	DR (1–0.52)	DR (0.67–0.52)	DR (0.52–0.27)	V_a m/s	V_w m/s		
0.54	0.25	83	85	81	3.24	1.4	94.6	96.3
0.62	0.47	87	83	69	3.24	2.45	85.2	94.4
0.5	0.47	85	79	81	3.24	3.49	83.0	87.3
0.66	0.25	82	89	87	5.97	2.45	83.6	88.0
0.62	0.35	87	82	73	5.97	3.49	80.7	81.2

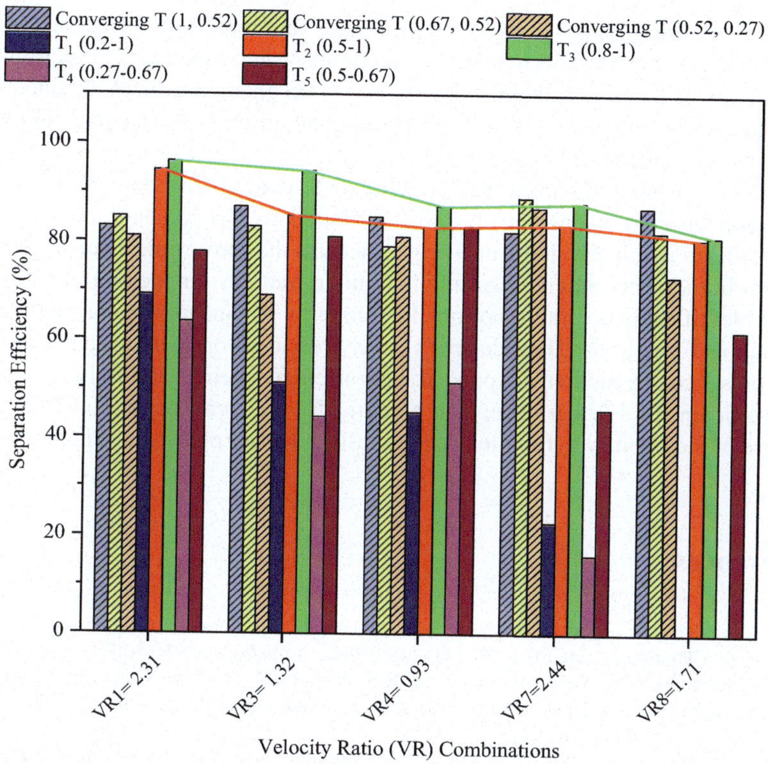

Fig. 3.42 Comparison of suddenly expanded T-junctions (T_{1-5}) with experimentation data of converging T-junction published by Memon [67]

It is of significant importance to highlight that Memon's experimental findings demonstrated lower separation efficiencies in comparison to the numerical simulation results presented by Pao and Memon [4] for converging T-junctions. Given that experimental data is derived from practical scenarios in the real world, it is imperative to engage in a comprehensive discussion and conduct a thorough comparison with the results obtained from experimentation.

Upon closer examination of T_2 and T_3, it becomes evident that they exhibit exceptional separation efficiencies of 96.3% and 94.6%, respectively, in contrast to the 89% obtained by the three converging T-junction arrangements. This results in a notable advantage, as T_3 exhibits a significant 7% improvement and T_2 demonstrates a 5% improvement compared to the converging T-junction competitors. Significantly, the separation efficiencies of T_2 and T_3 consistently remain above 80%, whereas the converging T-junctions (1–0.52), (0.67–0.52), and (0.52–0.27) exhibit decreases to 82%, 79%, and 69%, respectively.

An additional salient feature pertains to the enhanced efficacy exhibited by suddenly expanded T-junctions when subjected to elevated air and water superficial velocities. It is worth mentioning that in the conducted experiments, T-junctions with converging geometries were tested under the conditions of a maximum air superficial velocity of 0.66 m/s and a maximum water superficial velocity of 0.47 m/s. However, T-junctions with suddenly expanded geometries demonstrated higher efficiencies when subjected to considerably higher air (5.97 m/s) and water (3.49 m/s) superficial velocities.

In brief, it can be observed that T_1, T_4, and T_5 demonstrate comparatively lower separation efficiencies, which may be attributed to the presence of an abruptly expanded part within the side arm. The geometry of the converging T-junction enables efficient liquid containment, resulting in liquid retraction. On the other hand, T_2 and T_3 are identified as favorable options for suddenly expanded, regular, reduced, and converging T-junctions. The side arm's lower pressure drops, which can be attributed to the presence of a suddenly expanded section, enable them to attain significant separation efficiencies. Furthermore, their potential for practical applications is expanded because of their ability to function successfully at higher phase superficial velocities.

References

1. L. Yang, B.J. Azzopardi, A. Belghazi, S. Nakanishi, Phase separation of liquid-liquid two-phase flow at a T-junction. AIChE J. **52**, 141–149 (2006). https://doi.org/10.1002/aic.10589
2. M. Mohammadikharkeshi, R. Dabirian, R.S. Mohan, O. Shoham, Experimental study and modeling of slug dissipation in a horizontal enlarged impacting tee-junction. SPE J., 22–26 (2020). https://doi.org/10.2118/190131-pa
3. F. Ejaz, W. Pao, M. Shakir, A. Saieed, Z.Q. Memon, A review : Evolution of branching T-junction geometry in terms of diameter ratio, to improve phase separation Eng. Sci. Technol. an Int. J. (2021). https://doi.org/10.1016/j.jestch.2021.02.003
4. W. Pao, Z.Q. Memon, Numerical evaluation of separation efficiency in converging T-junction with slug flow. Int. J. Numer. Methods Heat Fluid Flow. **30**, 3515–3534 (2019). https://doi.org/10.1108/HFF-07-2019-0524
5. M. Tran, Z. Memon, A. Saieed, W. Pao, F. Hashim, Numerical simulation of two-phase separation in T-junction with experimental validation. J. Mech. Eng. Sci. **12**, 4216–4230 (2018). https://doi.org/10.15282/jmes.12.4.2018.17.0363
6. E. Wren, Geometric effects on phase split at a large diameter T-junction. Phd. Diss. (2001)
7. A. Katsaounis, G.F. Schultheiss, Diverging two-phase flow patterns in a T-junction and their influence on the form-resistance pressure drop (1985)
8. B.J. Azzopardi, D.A. Colman, D. Nicholson, Plant application of a T-junction as a partial phase separator. Chem. Eng. Res. Des. **80**, 87–96 (2002). https://doi.org/10.1205/026387602753393394
9. J. Chen, S. Wang, H. Ke, S. Cai, Y. Zhao, Gas-liquid two-phase flow splitting at microchannel junctions with different branch angles. Chem. Eng. Sci. **104**, 881–890 (2013). https://doi.org/10.1016/j.ces.2013.10.013
10. K.C. Hong, Two-phase flow splitting at a pipe tee. JPT, J. Pet. Technol. **30**, 290–296 (1978). https://doi.org/10.2118/6530-PA
11. S. Marti, O. Shoham, A unified model for stratified-wavy two-phase flow splitting at a reduced T-junction with an inclined branch arm. Int. J. Multiph. Flow **23**, 725–748 (1997). https://doi.org/10.1016/S0301-9322(97)82477-9
12. V.R. Penmatcha, P.J. Ashton, O. Shoham, Two-phase stratified flow splitting at a T-jun. Int. J. Multiph. Flow **22**, 1105–1122 (1996). https://doi.org/10.1016/0301-9322(96)00033-x
13. B.J. Azzopardi, P.A. Smith, Two-phase flow split at T junctions: effect of side arm orientation and downstream geometry. Int. J. Multiph. Flow **18**, 861–875 (1992)

14. W. Seeger, J. Reimann, U. Müller, Two-phase flow in a T-junction with a horizontal inlet. Part I: Phase separation. Int. J. Multiph. Flow. **12**, 575–585 (1986). https://doi.org/10.1016/0301-9322(86)90061-3
15. M. Ottens, H.C.J. Hoefsloot, P.J. Hamersma, Effect of small branch inclination on gas–liquid flow separation in T junctions. AIChE J. **45**, 465–474 (1999). https://doi.org/10.1002/aic.690450304
16. A. Azzi, A. Al-Attiyah, L. Qi, W. Cheema, B.J. Azzopardi, Gas-liquid two-phase flow division at a micro-T-junction. Chem. Eng. Sci. **65**, 3986–3993 (2010). https://doi.org/10.1016/j.ces.2010.03.037
17. G. Das, P.K. Das, B.J. Azzopardi, The split of stratified gas-liquid flow at a small diameter T-junction. Int. J. Multiph. Flow **31**, 514–528 (2005). https://doi.org/10.1016/j.ijmultiphaseflow.2005.01.009
18. B.J. Azzopardi, S. Rea, Phase separation using a simple T-junction. Proc. SPE Annu. Tech. Conf. Exhib. PI, 355–362 (2000). https://doi.org/10.2523/63040-ms
19. F. Peng, M. Shoukri, A.M.C. Chan, Effect of branch orientation on annular two-phase flow in T-Junctions. J. Fluids Eng. Trans. ASME. **118**, 166–171 (1996). https://doi.org/10.1115/1.2817496
20. N. Saba, R.T. Lahey, The analysis of phase separation phenomena in branching conduits. Int. J. Multiph. Flow **10**, 1–20 (1983). https://doi.org/10.1016/0301-9322(83)90057-5
21. J. Reimann, W. Seeger, Two-phase flow in a T-junction with a horizontal inlet. Part II: Pressure differences. Int. J. Multiph. Flow. **12**, 587–608 (1986). https://doi.org/10.1016/0301-9322(86)90062-5
22. S. Rea, B.J. Azzopardi, The split of horizontal stratified flow at a large diameter T-junction. Chem. Eng. Res. Des. **79**, 470–476 (2001). https://doi.org/10.1205/026387601750282409
23. S. Pandey, A. Gupta, D.P. Chakrabarti, G. Das, S. Ray, Liquid-liquid two phase flow through a horizontal T-junction. Chem. Eng. Res. Des. **84**, 895–904 (2006). https://doi.org/10.1205/cherd05061
24. A. Van Gorp, *Two-phase pressure drop and phase distribution at a reduced horizontal tee junction: the effect of system pressure* (University of Manitoba, Winnipeg, Manitoba, 1988)
25. A. Saieed, W. Pao, F.M. Hashim, Effect of T-junction diameter ratio on stratified-wavy flow separation. J. Nat. Gas Sci. Eng. **51**, 223–232 (2018). https://doi.org/10.1016/j.jngse.2018.01.015
26. E. Wren, G. Baker, B.J. Azzopardi, R. Jones, Slug flow in small diameter pipes and T-junctions. Exp. Therm. Fluid Sci. **29**, 893–899 (2005). https://doi.org/10.1016/j.expthermflusci.2005.01.004
27. A. Saieed, W. Pao, B. Hewakandamby, B.J. Azzopardi, D.A. Wood, H.M. Ali, Experimental investigation on the effect of diameter ratio on two-phase slug flow separation in a T-Junction. J. Pet. Sci. Eng. **170**, 139–150 (2018). https://doi.org/10.1016/j.petrol.2018.06.033
28. B.J. Azzopardi, P.B. Whalley, The effect of flow patterns on two-phase flow in a T junction. Int. J. Multiph. Flow **8**, 491–507 (1982). https://doi.org/10.1016/0301-9322(82)90020-9
29. B.J. Azzopardi, G. Freeman-Bell, The effect of the side arm diameter on the two-phase flow split at a "T" junction **10**, 509–512 (1983)
30. O. Shoham, J.P. Brill, Y. Taitel, Two-phase flow splitting in a tee junction-experiment and modelling. Chem. Eng. Sci. **42**, 2667–2676 (1987). https://doi.org/10.1016/0009-2509(87)87017-3
31. O. Shoham, S. Arirachakaran, J.P. Brill, Two-phase flow splitting in a horizontal reduced pipe tee. Chem. Eng. Sci. **44**, 2388–2391 (1989). https://doi.org/10.1016/0009-2509(89)85174-7
32. S.T. Hwang, H.M. Soliman, R.T. Lahey, Phase separation in dividing two-phase flows. Am. Inst. Chem. Eng. Work. **14**, 439–458 (1988)
33. J. Hart, P.J. Hamersma, J.M.H. Fortuin, A model for predicting liquid route preference during gas-liquid flow through horizontal branched pipelines. Chem. Eng. Sci. **46**, 1609–1622 (1991). https://doi.org/10.1016/0009-2509(91)87009-2
34. R.F. Mudde, J.S. Groen, H.E.A.V.D. Akker, Two-phase flow redistribution p h e n o m e n a in a large t-junction. Int. J. Multiph. Flow **19**, 563–573 (1993)

35. S. Griston, J.H. Choi, Two-Phase Flow Splitting at Side- B ranching Tees, 1–10 (1998)
36. J. Riemann, H.J. Brinkmann, R. Domanski, Gas-Liquid flow in dividing Tee-Junctions with a horizontal Inlet and different branch orientations and diameters, 187–229 (1988). 4Z72.70-4399
37. Casey, Two-phase pressure drop and phase distribution at reduced horizontal T-Junction: the effect of system pressure. MSc. Diss. **24**, 775–792 (1998). https://doi.org/10.1016/S0301-9322(98)00002-0
38. J.R. Buel, H.M. Soliman, G.E. Sims, Two-phase pressure drop and phase distribution of a horizontal tee junction. Int. J. Multiph. Flow **20**, 819–836 (1994). https://doi.org/10.1016/0301-9322(94)90096-5
39. L.C. Walters, Two-phase pressure drop and phase distribution at horizontal T-junction: the effect of branch diameter. University of Manitoba (1994)
40. J. Hart, P.J. Hamersma, J.M.H. Fortuin, Phase distribution during gas-liquid flow through horizontal dividing junctions. Nucl. Eng. Des. **126**, 293–312 (1991). https://doi.org/10.1016/0029-5493(91)90022-A
41. S. Rea, B.J. Azzopardi, The split of two-phase flow at a horizontal T-annular and stratified flow **79**, 3441–3453 (1989)
42. B.J. Azzopardi, Effect of side arm diameter on phase split at T-junctions. Proc. SPE Annu. Tech. Conf. Exhib. **2** (1999). https://doi.org/10.2523/56707-ms
43. W. Pao, A. Saieed, F.M. Hashim, R.B.M. Norpiah, Numerical simulation of two–phase separation in a small diameter ratio T-junction. Mat.-Wiss. u. Werkstofftech. **48**, 255–260 (2017). https://doi.org/10.1002/mawe.201600769
44. A. Saieed, *Experimental investigation on the effect of diameter ratio on two-phase separation in a T-junction* (Universiti Teknologi Petronas, Malaysia, 2018)
45. W. Pao, B. Sam, A. Saieed, C.M. Tran, Numerical investigation of liquid carryover in T-junction with different diameter ratios. IOP Conf. Ser. Mater. Sci. Eng. **328**, 1–6 (2018). https://doi.org/10.1088/1757-899X/328/1/012007
46. T.M.N. Tengku-Bazli, W. Pao, Experimental analyses on T-junction geometry in two phase flow (2019). https://doi.org/10.13140/RG.2.2.19357.97762.
47. S.A. Abd-Hadi, W. Pao, Experimental investigation of liquid carryover in modified T-junction, 2–8 (2019). https://doi.org/10.13140/RG.2.2.26068.86406
48. A. Saieed, Z.Q. Memon, M.C. Tran, W.P.K. Soon, A multiphase flow loop design for investigating the partial phase separation in a pipe tee. Acta Polytech. **59**, 399–410 (2019). https://doi.org/10.14311/AP.2019.59.0399
49. J. Reimann, W. Seeger, G3/GI equal 12, 587–608 (1986)
50. E. Wren, B.J. Azzopardi, The phase separation capabilities of two T-junctions placed in series. Chem. Eng. Res. Des. **82**, 364–371 (2004). https://doi.org/10.1205/026387604322870471
51. S.J. Tae, K. Cho, Two-phase flow distribution and phase separation through both horizontal and vertical branches. KSME Int. J. **17**, 1211–1218 (2003). https://doi.org/10.1007/BF03016516
52. Z. Memon, W. Pao, F. Hashim, S. Ahmed, Experimental investigation of two-phase separation in a horizontal T-junction with vertical branch arm. J. Mech. Eng. Sci. **13**, 5922–5937 (2019). https://doi.org/10.15282/jmes.13.4.2019.14.0470
53. Z.Q. Memon, W. Pao, F.M. Hashim, H.M. Ali, Experimental investigation of two-phase separation in T-Junction with combined diameter ratio. J. Nat. Gas Sci. Eng. **73**, 103048 (2020). https://doi.org/10.1016/j.jngse.2019.103048
54. M.N. Azad, W. Pao, CFD Simulation of Liquid Carryover in Hybrid Configuration of (2019). https://doi.org/10.13140/RG.2.2.23984.81920
55. D. Butterworth, Unsolved problems in heat exchanger design, in: Interflow 80—Fluid Handl. Conf., Harrogate, England (1980)
56. A.E. Fouda, E. Rhodes, Two-phase annular flow stream division in a simple TEE. Chem. Eng. Res. Des. **52a**, 354–360 (1974)
57. D.P. Margaris, T-junction separation modelling in gas-liquid two-phase flow. Chem. Eng. Process. Process Intensif. **46**, 150–158 (2007). https://doi.org/10.1016/j.cep.2006.05.007
58. C.T. Crowe, F. Group, Multiphase Flow Handbook (2006)

59. W. Pao, F.M. Hashim, L.H. Ming, Computational analyses of passive wet gas separation in branched piping. MATEC Web Conf. **13**, 1–5 (2014). https://doi.org/10.1051/matecconf/20141303009
60. J. Reimann, A new gas separator for stratified and slug flow.pdf, in: 3rd Int. Conf. Multiph. Flow, BHRA, The Hague, The Netherlands (1987)
61. M. Bevilacqua, G. Giacchetta, K. Neilson, Two-phase liquid-gas separation with a new conception comb separator, Bologna (2000)
62. G. H. Priestman, J.R. Tippetts, A compact gas-liquid separator with fluidic control, in: Trans. Inst. Chem. Eng.,pp. 690–697, Vol. 78A (2000)
63. L. Yang, Z. Zhao, P. Qi, L. Zhao, B.J. Azzopardi, Phase separation of gas–liquid two-phase flows in multitube T-junction separators, pp. 2285–2292 (2010)
64. L. Yang, J. Wang, B.J. Azzopardi, Z. Zhao, S. Xu, H. Wang, Phase Phase separation of gas–liquid two-phase stratified and plug flows in multitube T-junction separators. Am. Inst. Chem. Eng. J. 61, 1–15 (2014). https://doi.org/10.1002/aic.15552
65. J.L. Chen, L.M. He, X.M. Luo, H.T. Bai, Y.H. Wei, Simulation of oil-water two phase flow and separation behaviors in combined T junctions. J. Hydrodyn. **24**, 848–857 (2012). https://doi.org/10.1016/S1001-6058(11)60312-0
66. J. Zheng, K. Ou, H. Bie, P. Xu, Y. Zhao, X. Liu, Y. He, Heat transfer analysis of high-pressure hydrogen storage tanks subjected to localized firsse. Int. J. Hydrogen Energy. **37**, 13125–13131 (2012). https://doi.org/10.1016/j.ijhydene.2012.04.058
67. Q. Zeeshan, *Experimental Investigation of Two-Phase Separation in T-Junction with Combined Diameter Ratio*. Universiti Teknologi Petronas (2019)

Chapter 4
Conclusion and Recommendations

In conclusion, the extensive literature review underscores the ongoing issue of significant liquid carryover through the side arm outlet in T-junctions, despite their prevalent use as partial phase separators in various industrial contexts. The T-junction is a commonly chosen option due to its uncomplicated design, adaptability, and economical nature. However, various factors impact its effectiveness, particularly in situations involving two-phase slug flow regimes.

4.1 Conclusions

From the analysis of various studies, several key takeaways emerge:

(a) **Performance Assessment**: The literature presents a variety of criteria for assessing the performance of T-junctions. Various researchers utilize techniques such as gas and liquid fraction differentials, programmable logic controllers (PLCs), and liquid carryover thresholds. However, an effective and reliable approach involves the utilization of a method that calculates the dimensionless parameter "S" by integrating the area under the curve. However, it is essential to consider both peak liquid carryover and liquid carryover threshold as separate factors.

(b) **Diameter Ratio Impact**: The influence of diameter ratio on phase separation in T-junctions has been a subject of scholarly discourse. The enhancement of phase separation is typically observed when reducing the diameter ratio. However, it is important to note that there is a critical limit below which there is a significant deterioration in liquid carryover. A diameter ratio of approximately 0.67 has been determined to be advantageous for the phase separation of slug flow.

(c) **Side Arm Inclination**: The inclination of the side arm is a crucial factor in the redistribution of phases. Even a minor inclination can have an impact on the phenomenon of pressure drop, which in turn can have consequences on the

F. Ejaz et al., *Two Phase Separation in the T-Junction*,
https://doi.org/10.1007/978-981-99-9354-3_4

process of liquid extraction. It is advisable to position the side arm in a vertical upward orientation in order to enhance the process of phase separation.

(d) **Superficial Velocity Effect**: The effects of superficial velocity on phase redistribution are significant, particularly in relation to liquid and air flow rates. Phase separation is enhanced by higher liquid superficial velocities as they counteract the effects of inertia. On the other hand, increased air velocities result in pressure drop and encourage the diversion of liquid.

Despite the extensive corpus of research, there remains a persistent need for further improvement in T-junction design. The investigation into the modification of side arm diameter ratios is currently lacking in scholarly attention. The study highlights the difficulty of achieving a proper equilibrium between peak liquid carryover and the liquid carryover threshold in both conventional and modified T-junction designs, thus calling for the adoption of a novel methodology.

In light of the aforementioned challenges, a researcher presented a novel design concept: an abruptly enlarged T-junction incorporating a side arm characterized by a sudden expansion in its profile. The primary objective of this novel design was to enhance the efficiency of separation while simultaneously preserving the essential performance indicators of LCT and PLC.

Their findings provide insights into the behavior of various suddenly expanded T-junction designs across different conditions.

(a) **Liquid Superficial Velocity**: The experimental results indicated that higher liquid superficial velocities led to a decrease in separation efficiency, a slight decrease in liquid carryover time (LCT), and a significant increase in pressure loss coefficient (PLC) for all five suddenly expanded T-junction configurations. This behavior is consistent with concepts found in the existing literature.

(b) **Air Superficial Velocity**: The impact of increased air velocities on separation efficiency, particulate loading capacity (PLC), and pressure drop (LCT) was investigated, confirming existing theoretical principles.

(c) **Diameter Ratio Effects**: The effects of diameter ratio on separation efficiency, pressure loss coefficient (PLC), and liquid carryover (LCT) were found to be improved when the upstream diameter ratio (DR1) values were equal to or greater than 0.5. It is worth mentioning that a downstream diameter ratio (DR2) of 1 exhibited enhanced phase separation in comparison to a ratio of 0.67. The critical values for diameter ratios were determined to be DR1 $\geq$ 0.5 and DR2 = 1. These values indicate that diameter ratios below these thresholds should be avoided in order to guarantee sufficient phase separation.

(d) **Performance Improvement**: In the context of T-junction designs that have experienced sudden expansion, it was observed that T2 and T3 demonstrated significant improvements in separation efficiency. Specifically, T2 exhibited an enhancement of 11%, while T3 showed a 13% improvement. These results were compared to previously published simulation and experimental data for converging T-junctions. The T-junctions exhibited exceptional separation efficiencies of 96.3 and 94.6%, surpassing the performance of conventional designs.

(e) **Overall Efficiency**: The study revealed that the sudden expansion T-junction designs consistently achieved a separation efficiency of no less than 80.7%, indicating their reliable and consistent performance. On the other hand, the T-junction configuration with the lowest efficiency exhibited a separation efficiency of only 69%.

In conclusion, the study highlights the potential of abruptly expanded T-junctions as a viable design alternative for improving the efficiency of separation in air-water slug flow. The results underscore the significance of meticulously chosen diameter ratios and offer valuable insights into the performance benefits afforded by these innovative configurations. This study provides significant contributions to the discipline of fluid dynamics and offers opportunities for further investigation and enhancement of T-junction designs in order to fulfill the phase separation needs of various industries.

4.2 Recommendations

Based on the findings and scope of the research study, several noteworthy recommendations for future investigation in this domain can be delineated:

(a) **Experimental Validation**: In order to enhance the reliability of the recently proposed suddenly expanded T-junction design, it would be appropriate to proceed with experimental investigations focused on evaluating the separation efficiency. Empirical evidence of the design's performance and validation of its phase separation capabilities can be obtained through real-world testing.

(b) **Varied Working Fluids**: The inclusion of diverse working fluids, such as oil and gas, can broaden the scope of the study and improve its applicability and resilience. The implementation of a diversified approach would enhance the comprehensiveness of the assessment of the effectiveness of the T-junction design in different fluid dynamics scenarios.

(c) **Data Consolidation and Deep Learning**: The consolidation of data and the application of deep learning techniques can prove beneficial in the context of abundant and dispersed experimental and numerical simulation data. It is advantageous to undertake a systematic endeavor to collect and consolidate this information. The utilization of deep-learning neural networks has the potential to facilitate the detection of ideal T-junction configurations suitable for various flow conditions. The aforementioned approach exhibits the capacity to optimize the design process and reduce the necessity for extensive experimentation.

(d) **Dynamic Prediction Model**: The development of a dynamic prediction model represents a novel and underexplored approach. By utilizing a statistical methodology and algorithmic equations that are capable of adapting to the unique characteristics of the application, a predictive model has the potential to dynamically modify itself in response to varying conditions. This methodology possesses the capacity to substantially diminish the duration and expenses associated with repetitive experimentation and data acquisition.

(e) **Practical Implementation Studies**: The practical implementation of studies involves establishing collaborations with industrial partners to apply the innovative T-junction design in real-world scenarios, thereby facilitating the integration of theoretical research with practical utility. Case studies that involve real-world industrial processes can serve as effective demonstrations of the efficacy of a particular design in enhancing phase separation.

(f) **Environmental Considerations**: The examination of the environmental effects associated with the recently proposed T-junction design, including its capacity to decrease energy usage and emissions, has the potential to introduce a sustainability aspect to the study.

The recommendations not only serve to expand upon existing research but also present intriguing prospects for furthering the domain of T-junction design and phase separation. Every recommendation possesses the capacity to contribute towards increased efficiency, decreased expenses, and enhanced industrial procedures.

Appendix

See Tables 1, 2, 3, 4, 5 and 6.

F. Ejaz et al., *Two Phase Separation in the T-Junction*,
https://doi.org/10.1007/978-981-99-9354-3

Table 1 Numerical simulation results for five suddenly expanded T-junctions with slug flow

Sr. no	M_{SR}	DR_1	DR_2	VR	F_a	F_w
1	0.2	0.2	1	2.314	0.208572	0
2	0.4	0.2	1	2.314	0.526245	0.0533811
3	0.5	0.2	1	2.314	0.637691	0.0946435
4	0.6	0.2	1	2.314	0.716359	0.173772
5	0.8	0.2	1	2.314	0.994118	0.74174
6	0.2	0.5	1	2.314	0.303235	0
7	0.4	0.5	1	2.314	0.670588	0
8	0.5	0.5	1	2.314	0.763971	0.0506357
9	0.6	0.5	1	2.314	0.814647	0.0897509
10	0.8	0.5	1	2.314	0.952941	0.218495
11	0.2	0.8	1	2.314	0.292896	0
12	0.4	0.8	1	2.314	0.673843	0
13	0.5	0.8	1	2.314	0.704943	0.029371
14	0.6	0.8	1	2.314	0.829065	0.0465188
15	0.8	0.8	1	2.314	0.953028	0.165639
16	0.2	0.27	0.67	2.314	0.273399	0
17	0.4	0.27	0.67	2.314	0.541723	0.100253
18	0.5	0.27	0.67	2.314	0.651489	0.183332
19	0.6	0.27	0.67	2.314	0.744342	0.304617
20	0.8	0.27	0.67	2.314	0.989967	0.755528
21	0.2	0.5	0.67	2.314	0.282328	0
22	0.4	0.5	0.67	2.314	0.545263	0
23	0.5	0.5	0.67	2.314	0.6745	0.131287
24	0.6	0.5	0.67	2.314	0.793595	0.193222
25	0.8	0.5	0.67	2.314	0.989967	0.652161
26	0.2	0.2	1	1.851	0.295105	0
27	0.4	0.2	1	1.851	0.563975	0.120081
28	0.5	0.2	1	1.851	0.657601	0.235726
29	0.6	0.2	1	1.851	0.800558	0.651122
30	0.8	0.2	1	1.851	0.988899	0.805839
31	0.2	0.5	1	1.851	0.303646	0
32	0.4	0.5	1	1.851	0.592635	0
33	0.5	0.5	1	1.851	0.731459	0.102256
34	0.6	0.5	1	1.851	0.835836	0.13655
35	0.8	0.5	1	1.851	0.962297	0.35984
36	0.2	0.8	1	1.851	0.31522	0
37	0.4	0.8	1	1.851	0.613269	0

(continued)

Table 1 (continued)

Sr. no	M_{SR}	DR_1	DR_2	VR	F_a	F_w
38	0.5	0.8	1	1.851	0.753592	0.0623318
39	0.6	0.8	1	1.851	0.878585	0.0996705
40	0.8	0.8	1	1.851	0.96223	0.343388
41	0.2	0.27	0.67	1.851	0.292961	0
42	0.4	0.27	0.67	1.851	0.584537	0.0827675
43	0.5	0.27	0.67	1.851	0.69429	0.169211
44	0.6	0.27	0.67	1.851	0.738489	0.432496
45	0.8	0.27	0.67	1.851	0.961489	0.823082
46	0.2	0.5	0.67	1.851	0.303295	0
47	0.4	0.5	0.67	1.851	0.586157	0
48	0.5	0.5	0.67	1.851	0.727182	0.109812
49	0.6	0.5	0.67	1.851	0.832287	0.204675
50	0.8	0.5	0.67	1.851	0.961506	0.66951
51	0.2	0.2	1	1.322	0.335444	0
52	0.4	0.2	1	1.322	0.554086	0.180954
53	0.5	0.2	1	1.322	0.724431	0.497314
54	0.6	0.2	1	1.322	0.886538	0.780598
55	0.8	0.2	1	1.322	0.964465	0.837465
56	0.2	0.5	1	1.322	0.345684	0
57	0.4	0.5	1	1.322	0.716618	0
58	0.5	0.5	1	1.322	0.754347	0.176477
59	0.6	0.5	1	1.322	0.840625	0.306801
60	0.8	0.5	1	1.322	0.963798	0.50151
61	0.2	0.8	1	1.322	0.351893	0
62	0.4	0.8	1	1.322	0.776563	0
63	0.5	0.8	1	1.322	0.822094	0.0936044
64	0.6	0.8	1	1.322	0.92344	0.189477
65	0.8	0.8	1	1.322	0.964015	0.374039
66	0.2	0.27	0.67	1.322	0.345684	0
67	0.4	0.27	0.67	1.322	0.443269	0.159637
68	0.5	0.27	0.67	1.322	0.530544	0.269471
69	0.6	0.27	0.67	1.322	0.635319	0.451621
70	0.8	0.27	0.67	1.322	0.962297	0.858887
71	0.2	0.5	0.67	1.322	0.343732	0
72	0.4	0.5	0.67	1.322	0.606999	0
73	0.5	0.5	0.67	1.322	0.7563	0.173947
74	0.6	0.5	0.67	1.322	0.836555	0.295637

(continued)

Table 1 (continued)

Sr. no	M_{SR}	DR_1	DR_2	VR	F_a	F_w
75	0.8	0.5	0.67	1.322	0.963779	0.70197
76	0.2	0.2	1	0.9281	0.380929	0
77	0.4	0.2	1	0.9281	0.485008	0.196226
78	0.5	0.2	1	0.9281	0.59512	0.372956
79	0.6	0.2	1	0.9281	0.814518	0.688114
80	0.8	0.2	1	0.9281	0.964421	0.860381
81	0.2	0.5	1	0.9281	0.407564	0
82	0.4	0.5	1	0.9281	0.674105	0
83	0.5	0.5	1	0.9281	0.741565	0.195417
84	0.6	0.5	1	0.9281	0.814518	0.284778
85	0.8	0.5	1	0.9281	0.961556	0.545951
86	0.2	0.8	1	0.9281	0.41132	0
87	0.4	0.8	1	0.9281	0.666495	0
88	0.5	0.8	1	0.9281	0.706833	0.116468
89	0.6	0.8	1	0.9281	0.903632	0.268831
90	0.8	0.8	1	0.9281	0.962001	0.534359
91	0.2	0.27	0.67	0.9281	0.389082	0
92	0.4	0.27	0.67	0.9281	0.620672	0.288497
93	0.5	0.27	0.67	0.9281	0.695169	0.464987
94	0.6	0.27	0.67	0.9281	0.79943	0.629324
95	0.8	0.27	0.67	0.9281	0.964427	0.888163
96	0.2	0.5	0.67	0.9281	0.40823	0
97	0.4	0.5	0.67	0.9281	0.593155	0
98	0.5	0.5	0.67	0.9281	0.823173	0.216144
99	0.6	0.5	0.67	0.9281	0.874809	0.361302
100	0.8	0.5	0.67	0.9281	0.955624	0.723973
101	0.2	0.2	1	4.2678	0.149533	0
102	0.4	0.2	1	4.2678	0.441334	0.223463
103	0.5	0.2	1	4.2678	0.55224	0.27674
104	0.6	0.2	1	4.2678	0.588706	0.484556
105	0.8	0.2	1	4.2678	0.990238	0.902278
106	0.2	0.5	1	4.2678	0.109883	0
107	0.4	0.5	1	4.2678	0.51094	0
108	0.5	0.5	1	4.2678	0.821572	0.176626
109	0.6	0.5	1	4.2678	0.921087	0.308021
110	0.8	0.5	1	4.2678	0.963901	0.593993
111	0.2	0.8	1	4.2678	0.238352	0

(continued)

Table 1 (continued)

Sr. no	M_{SR}	DR_1	DR_2	VR	F_a	F_w
112	0.4	0.8	1	4.2678	0.556002	0
113	0.5	0.8	1	4.2678	0.852364	0.124142
114	0.6	0.8	1	4.2678	0.896652	0.197509
115	0.8	0.8	1	4.2678	0.975236	0.502932
116	0.2	0.27	0.67	4.2678	0.231422	0
117	0.4	0.27	0.67	4.2678	0.451299	0.180896
118	0.5	0.27	0.67	4.2678	0.55418	0.268424
119	0.6	0.27	0.67	4.2678	0.628845	0.476379
120	0.8	0.27	0.67	4.2678	0.974813	0.918986
121	0.2	0.5	0.67	4.2678	0.236468	0
122	0.4	0.5	0.67	4.2678	0.465705	0
123	0.5	0.5	0.67	4.2678	0.575475	0.177534
124	0.6	0.5	0.67	4.2678	0.671128	0.417721
125	0.8	0.5	0.67	4.2678	0.970898	0.785316
126	0.2	0.2	1	3.4	0.157162	0
127	0.4	0.2	1	3.4	0.427228	0.306929
128	0.5	0.2	1	3.4	0.655883	0.634744
129	0.6	0.2	1	3.4	0.751464	0.774573
130	0.8	0.2	1	3.4	0.929005	0.903802
131	0.2	0.5	1	3.4	0.250507	0
132	0.4	0.5	1	3.4	0.55652	0
133	0.5	0.5	1	3.4	0.855481	0.177263
134	0.6	0.5	1	3.4	0.898572	0.305837
135	0.8	0.5	1	3.4	0.971102	0.699596
136	0.2	0.8	1	3.4	0.248964	0
137	0.4	0.8	1	3.4	0.581464	0
138	0.5	0.8	1	3.4	0.805025	0.11807
139	0.6	0.8	1	3.4	0.892509	0.245716
140	0.8	0.8	1	3.4	0.950828	0.540289
141	0.2	0.27	0.67	3.4	0.242973	0
142	0.4	0.27	0.67	3.4	0.461112	0.191205
143	0.5	0.27	0.67	3.4	0.663963	0.575015
144	0.6	0.27	0.67	3.4	0.758602	0.843976
145	0.8	0.27	0.67	3.4	0.941934	0.931506
146	0.2	0.5	0.67	3.4	0.250015	0
147	0.4	0.5	0.67	3.4	0.487041	0
148	0.5	0.5	0.67	3.4	0.597875	0.165475

(continued)

Table 1 (continued)

Sr. no	M_{SR}	DR_1	DR_2	VR	F_a	F_w
149	0.6	0.5	0.67	3.4	0.685954	0.306089
150	0.8	0.5	0.67	3.4	0.952211	0.808073
151	0.2	0.2	1	2.438	0.213987	0
152	0.4	0.2	1	2.438	0.425787	0.228894
153	0.5	0.2	1	2.438	0.683415	0.737372
154	0.6	0.2	1	2.438	0.76696	0.796812
155	0.8	0.2	1	2.438	0.964425	0.956371
156	0.2	0.5	1	2.438	0.267832	0
157	0.4	0.5	1	2.438	0.51239	0
158	0.5	0.5	1	2.438	0.804453	0.186382
159	0.6	0.5	1	2.438	0.902544	0.290324
160	0.8	0.5	1	2.438	0.966017	0.712703
161	0.2	0.8	1	2.438	0.271248	0
162	0.4	0.8	1	2.438	0.525039	0
163	0.5	0.8	1	2.438	0.746558	0.0559435
164	0.6	0.8	1	2.438	0.903528	0.283311
165	0.8	0.8	1	2.438	0.963534	0.595476
166	0.2	0.27	0.67	2.438	0.251051	0
167	0.4	0.27	0.67	2.438	0.384948	0.1423
168	0.5	0.27	0.67	2.438	0.435827	0.323254
169	0.6	0.27	0.67	2.438	0.776124	0.856691
170	0.8	0.27	0.67	2.438	0.990758	0.977506
171	0.2	0.5	0.67	2.438	0.266728	0
172	0.4	0.5	0.67	2.438	0.511974	0
173	0.5	0.5	0.67	2.438	0.602417	0.236902
174	0.6	0.5	0.67	2.438	0.747424	0.698108
175	0.8	0.5	0.67	2.438	0.990758	0.879062
176	0.2	0.2	1	1.711	0.22102	0
177	0.4	0.2	1	1.711	0.389426	0.542643
178	0.5	0.2	1	1.711	0.56999	0.769692
179	0.6	0.2	1	1.711	0.795249	0.91179
180	0.8	0.2	1	1.711	0.987364	0.958734
181	0.2	0.5	1	1.711	0.28633	0
182	0.4	0.5	1	1.711	0.549922	0
183	0.5	0.5	1	1.711	0.847913	0.2432
184	0.6	0.5	1	1.711	0.900472	0.357876
185	0.8	0.5	1	1.711	0.979534	0.773105

(continued)

Table 1 (continued)

Sr. no	M_{SR}	DR_1	DR_2	VR	F_a	F_w
186	0.2	0.8	1	1.711	0.296857	0
187	0.4	0.8	1	1.711	0.548517	0
188	0.5	0.8	1	1.711	0.802557	0.255625
189	0.6	0.8	1	1.711	0.892935	0.361586
190	0.8	0.8	1	1.711	0.959262	0.689563
191	0.2	0.27	0.67	1.711	0.25719	0
192	0.4	0.27	0.67	1.711	0.331388	0.471079
193	0.5	0.27	0.67	1.711	0.464495	0.717745
194	0.6	0.27	0.67	1.711	0.791726	0.911949
195	0.8	0.27	0.67	1.711	0.975536	0.997226
196	0.2	0.5	0.67	1.711	0.288057	0
197	0.4	0.5	0.67	1.711	0.543454	0
198	0.5	0.5	0.67	1.711	0.709015	0.277303
199	0.6	0.5	0.67	1.711	0.796077	0.443905
200	0.8	0.5	0.67	1.711	0.99568	0.920437

Table 2 Numerical results in terms of Liquid Carryover Threshold (LCT), Peak Liquid Carryover (PL) and Separation Efficiency (SE)

Suddenly expanded T-junction no	Velocity ratio no	V_a (m/s)	V_w (m/s)	LCT	PLC	SE (%)
T_1	VR_1	3.24	1.4	0.20857	0.74174	69.114
	VR_2		1.75	0.2951	0.80584	53.322
	VR_3		2.45	0.33544	0.83747	51.164
	VR_4		3.491	0.38093	0.86038	45.198
	VR_5	5.975	1.4	0.14953	0.90228	29.484
	VR_6		1.75	0.15716	0.9038	26.914
	VR_7		2.45	0.21399	0.95637	22.84
	VR_8		3.491	0.22102	0.95873	0
T_2	VR_1	3.24	1.4	0.67059	0.2185	94.552
	VR_2		1.75	0.59264	0.35984	89.81
	VR_3		2.45	0.71662	0.50151	85.208
	VR_4		3.491	0.6741	0.54595	82.964
	VR_5	5.975	1.4	0.51094	0.59399	86.006

(continued)

Table 2 (continued)

Suddenly expanded T-junction no	Velocity ratio no	V_a (m/s)	V_w (m/s)	LCT	PLC	SE (%)
	VR_6		1.75	0.55652	0.6996	85.622
	VR_7		2.45	0.51239	0.7127	83.56
	VR_8		3.491	0.54992	0.77311	80.694
T_3	VR_1	3.24	1.4	0.67384	0.16564	96.336
	VR_2		1.75	0.61327	0.34339	93.394
	VR_3		2.45	0.77656	0.37404	94.418
	VR_4		3.491	0.66649	0.53436	87.26
	VR_5	5.975	1.4	0.556	0.50293	89.65
	VR_6		1.75	0.58146	0.54029	89.594
	VR_7		2.45	0.525	0.59548	87.992
	VR_8		3.491	0.54852	0.68956	81.168
T_4	VR_1	3.24	1.4	0.2734	0.75553	63.626
	VR_2		1.75	0.29296	0.82308	64.162
	VR_3		2.45	0.34568	0.85889	44.292
	VR_4		3.491	0.38908	0.88816	51.258
	VR_5	5.975	1.4	0.23142	0.91899	37.564
	VR_6		1.75	0.24297	0.93151	34.306
	VR_7		2.45	0.25105	0.97751	16.22
	VR_8		3.491	0.25719	0.99723	0
T_5	VR_1	3.24	1.4	0.54526	0.65216	77.84
	VR_2		1.75	0.58616	0.66951	83.86
	VR_3		2.45	0.607	0.70197	80.942
	VR_4		3.491	0.59316	0.72397	83.276
	VR_5	5.975	1.4	0.46571	0.78532	56.294
	VR_6		1.75	0.48704	0.80807	64.348
	VR_7		2.45	0.51197	0.87906	45.922
	VR_8		3.491	0.54345	0.92044	61.898

Table 3 Data set published by Memon et al. [1]

Sr. no	Test set run	T-Junction	Gas superficial velocity (m/s)	Liquid superficial velocity (m/s)	VR	Flow regime
1	TSR1	T1	0.427	0.159	2.686	Stratified-wavy
2	TSR2	T1	0.427	0.202	2.1 14	Stratified-wavy
3	TSR3	T1	0.427	0.252	1.694	Slug
4	TSR4	T1	0.543	0.159	3.415	Stratified-wavy
5	TSR5	T1	0.543	0.202	2.688	Stratified-wavy
6	TSR6	T1	0.543	0.252	2.155	Slug
7	TSR7	T1	0.659	0.159	4.145	Stratified-wavy
8	TSR8	T1	0.659	0.202	3.262	Stratified-wavy
9	TSR9	T1	0.659	0.252	2.615	Slug
10	TSR10	T1	0.388	0.349	1.112	Slug
11	TSR11	T1	0.388	0.465	0.834	Slug
12	TSR12	T1	0.388	0.582	0.667	Slug
13	TSR13	T1	0.504	0.349	1.444	Slug
14	TSR14	T1	0.504	0.465	1.084	Slug
15	TSR15	T1	0.504	0.582	0.866	Slug
16	TSR16	T1	0.62	0.349	1.777	Slug
17	TSR17	T1	0.62	0.465	1.333	Slug
18	TSR18	T1	0.62	0.582	1.065	Slug
19	TSR1	T2	0.427	0.159	2.686	Stratified-wavy
20	TSR2	T2	0.427	0.202	2.114	Stratified-wavy
21	TSR3	T2	0.427	0.252	1.694	Slug
22	TSR4	T2	0.543	0.159	3.415	Stratified-wavy
23	TSR5	T2	0.543	0.202	2.688	Stratified-wavy
24	TSR6	T2	0.543	0.252	2.155	Slug
25	TSR7	T2	0.659	0.159	4.145	Stratified-wavy
26	TSR8	T2	0.659	0.202	3.262	Stratified-wavy
27	TSR9	T2	0.659	0.252	2.615	Slug
28	TSR10	T2	0.388	0.349	1.112	Slug
29	TSR11	T2	0.388	0.465	0.834	Slug
30	TSR12	T2	0.388	0.582	0.667	Slug
31	TSR13	T2	0.504	0.349	1.444	Slug
32	TSR14	T2	0.504	0.465	1.084	Slug
33	TSR15	T2	0.504	0.582	0.866	Slug
34	TSR16	T2	0.62	0.349	1.777	Slug
35	TSR17	T2	0.62	0.465	1.333	Slug

(continued)

Table 3 (continued)

Sr. no	Test set run	T-Junction	Gas superficial velocity (m/s)	Liquid superficial velocity (m/s)	VR	Flow regime
36	TSR18	T2	0.62	0.582	1.065	Slug
37	TSRI	T3	0.427	0.159	2.686	Stratified-wavy
38	TSR2	T3	0.427	0.202	2.114	Stratified-wavy
39	TSR3	T3	0.427	0.252	1.694	Slug
40	TSR4	T3	0.543	0.159	3.415	Stratified-wavy
41	TSR5	T3	0.543	0.202	2.688	Stratified-wavy
42	TSR6	T3	0.543	0.252	2.155	Slug
43	TSR7	T3	0.659	0.159	4.145	Stratified-wavy
44	TSR8	T3	0.659	0.202	3.262	Stratified-wavy
45	TSR9	T3	0.659	0.252	2.615	Slug
46	TSR10	T3	0.388	0.349	1.112	Slug
47	TSR11	T3	0.388	0.465	0.834	Slug
48	TSR12	T3	0.388	0.582	0.667	Slug
49	TSR13	T3	0.504	0.349	1.444	Slug
50	TSR14	T3	0.504	0.465	1.084	Slug
51	TSR15	T3	0.504	0.582	0.866	Slug
52	TSR16	T3	0.62	0.349	1.777	Slug
53	TSR17	T3	0.62	0.465	1.333	Slug
54	TSR18	T3	0.62	0.582	1.065	Slug
55	TSR1	T4	0.427	0.159	2.686	Stratified-wavy
56	TSR2	T4	0.427	0.202	2.114	Stratified-wavy
57	TSR3	T4	0.427	0.252	1.694	Slug
58	TSR4	T4	0.543	0.159	3.415	Stratified-wavy
59	TSR5	T4	0.543	0.202	2.688	Stratified-wavy
60	TSR6	T4	0.543	0.252	2.155	Slug
61	TSR7	T4	0.659	0.159	4.145	Stratified-wavy
62	TSR8	T4	0.659	0.202	3.262	Stratified-wavy
63	TSR9	T4	0.659	0.252	2.615	Slug
64	TSR10	T4	0.388	0.349	1.112	Slug
65	TSR11	T4	0.388	0.465	0.834	Slug
66	TSR12	T4	0.388	0.582	0.667	Slug
67	TSR13	T4	0.504	0.349	1.444	Slug
68	TSR14	T4	0.504	0.465	1.084	Slug
69	TSR15	T4	0.504	0.582	0.866	Slug
70	TSR16	T4	0.62	0.349	1.777	Slug

(continued)

Table 3 (continued)

Sr. no	Test set run	T-Junction	Gas superficial velocity (m/s)	Liquid superficial velocity (m/s)	VR	Flow regime
71	TSR17	T4	0.62	0.465	1.333	Slug
72	TSR18	T4	0.62	0.582	1.065	Slug
73	TSR1	T5	0.427	0.159	2.686	Stratified-wavy
74	TSR2	T5	0.427	0.202	2.1 14	Stratified-wavy
75	TSR3	T5	0.427	0.252	1.694	Slug
76	TSR4	T5	0.543	0.159	3.415	Stratified-wavy
77	TSR5	T5	0.543	0.202	2.688	Stratified-wavy
78	TSR6	T5	0.543	0.252	2.155	Slug
79	TSR7	T5	0.659	0.159	4.145	Stratified-wavy
80	TSR8	T5	0.659	0.202	3.262	Stratified-wavy
81	TSR9	T5	0.659	0.252	2.615	Slug
82	TSR10	T5	0.388	0.349	1.1 12	Slug
83	TSR11	T5	0.388	0.465	0.834	Slug
84	TSR12	T5	0.388	0.582	0.667	Slug
85	TSR13	T5	0.504	0.349	1.444	Slug
86	TSR14	T5	0.504	0.465	1.084	Slug
87	TSR15	T5	0.504	0.582	0.866	Slug
88	TSR16	T5	0.62	0.349	1.777	Slug
89	TSR17	T5	0.62	0.465	1.333	Slug
90	TSR18	T5	0.62	0.582	1.065	Slug

Table 4 Data set published by Memon et al. [1] for LCT and PLC

DR	VCN	Flow regime	LCT	Peak liquid carryovers
1	1	Stratified-wavy	0.91	15% at 100% air extraction
1	2	Stratified-wavy	0.90	Extraction of air 14.4% at 100% air extraction
1	3	Stratified-wavy	0.90	Extraction of air 13% at 100% air extraction
1	4	Stratified-wavy	0.90.8	Extraction of air 20% at 100% air extraction
1	5	Stratified-wavy	0.89.6	19% at 100% extraction of air
1	6	Stratified-wavy	0.884	16.3% at 100% air extraction
1	7	Stratified-wavy	0.89	26% at 100% air extraction
1	8	Stratified-wavy	0.87	24% at 100% air extraction
1	9	Stratified-wavy	0.86	20% at 100% air extraction
1	10	Slug	0.79	83% at 96% air extraction
1	11	Slug	0.785	74% at 100% air extraction
1	12	Slug	0.77	66% at 98% air extraction
1	13	Slug	0.75	54% at 98% extraction of air
1	14	Slug	0.79	91% at 94% air extraction
1	15	Slug	0.778	79.4% at 98% air extraction
1	16	Slug	0.769	72% at 99% air extraction
1	17	Slug	0.73	60% at 100% air extraction
1	18	Slug	0.77	96% at 99% air extraction
1	19	Slug	0.75	92.7% at 98% air extraction
1	20	Slug	0.71	84% at 99% air extraction
0.7	1	Stratified-wavy	0.89	13% at 100% air extraction
0.7	2	Stratified-wavy	0.87	11% at 100% air extraction
0.7	3	Stratified-wavy	0.86	9% at 100% air extraction
0.7	4	Stratified-wavy	0.883	18% at 100% air extraction
0.7	5	Stratified-wavy	0.86	14% at 100% air extraction
0.7	6	Stratified-wavy	0.84.6	11.9% at 100% air extraction
0.7	7	Stratified-wavy	0.87	23% at 100% air extraction
0.7	8	Stratified-wavy	0.85.3	20.8% at 100% air extraction
0.7	9	Stratified-wavy	0.836	17.4% at 100% air extraction
0.7	10	Slug	0.75	61.75% at 98% air extraction
0.7	11	Slug	0.745	53% at 97% extraction of air
0.7	12	Slug	0.73	45% at 99% air extraction
0.7	13	Slug	0.69	39% at 100% air extraction
0.7	14	Slug	0.73	73% at 98.8% air extraction
0.7	15	Slug	0.72	62% at 97.6% air extraction
0.7	16	Slug	0.68	50% at 98% air extraction
0.7	17	Slug	0.67	42.5% at 99.3% air extraction

(continued)

Table 4 (continued)

DR	VCN	Flow regime	LCT	Peak liquid carryovers
0.7	18	Slug	0.705	83% at 100% air extraction
0.7	19	Slug	0.678	77% at 99% air extraction
0.7	20	Slug	0.65	66% at 98.6% air extraction
0.52	1	Stratified-wavy	0.859	11%at 100% air extraction
0.52	2	Stratified-wavy	0.84	10% at 100% air extraction
0.52	3	Stratified-yvavy	0.83	7% at 100% air extraction
0.52	4	Stratified-wavy	0.848	17% at 100% air extraction
0.52	5	Stratified-wavy	0.82	12.2% at 100% air extraction
0.52	6	Stratified-wavy	0.809	9% at 100% air extraction
0.52	7	Stratified-wavy	0.83	21% at 100% air extraction
0.52	8	Stratified-wavy	0.825	18.6% at 100% air extraction
0.52	9	Stratified-wavy	0.79	16.2% at 100% air extraction
0.52	10	Slug	0.71	55.3% at 99% air extraction
0.52	11	Slug	0.60	46% at 98.5%
0.52	12	Slug	0.52	35% at 98% extraction of air
0.52	13	Slug	0.52	31% at 99% air extraction
0.52	14	Slug	0.65	65% at 98.1% air extraction
0.52	15	Slug	0.53	55% at 97.9% extraction of air
0.52	16	Slug	0.47	46% at 99.5% air extraction
0.52	17	Slug	0.23	35% at 99% air extraction
0.52	18	Semi-slug	0.74	59.4% at 99% air extraction
0.52	19	Slug	0.41	69% at 99% air extraction
0.52	20	Slug	0.35	58% at 99% air extraction
0.3	1	Stratified-yvavy	0.79	8% at 100% air extraction
0.3	2	Stratified-wavy	0.75	6.9% at 100% air extraction
0.3	3	Stratified-wavy	0.70%	5.2% at 100% air extraction
0.3	4	Stratified-wavy	0.77	13% at 100% air extraction
0.3	5	Stratified-wavy	0.733	10.2% at 100% air extraction
0.3	6	Stratified-wavy	0.696	6.5% at 100% air extraction
0.3	7	Stratified-wavy	0.75	17% at 100% air extraction
0.3	8	Stratified-yvavy	0.712	15.7% at 100% air extraction
0.3	9	Stratified-wavy	0.673	12.1% at 100% air extraction
0.3	10	Slug	0.61	35% at 97% extraction of air
0.3	11	Slug	0.405	28% at 96% extraction of air
0.3	12	Slug	0.27	22% at 95% air extraction
0.3	13	Slug	0.26	18.6% at 96% air extraction
0.3	14	Slug	0.50	45.4% at 97.6% air extraction

(continued)

Table 4 (continued)

DR	VCN	Flow regime	LCT	Peak liquid carryovers
0.3	15	Slug	0.27	35% at 96.5% air extraction
0.3	16	Slug	0.24	28.2% at 97% air extraction
0.3	17	Siug	0.20	24% at 98% air extraction
0.3	18	Semi-slug	0.66	33% at 98.6% air extraction
0.3	19	Slug	0.38	52% at 97% air extraction
0.3	20	Slug	0.27	41% at 98.5% air extraction
0.2	1	Stratified-wavy	0.69.8	6% at 100% air extraction
0.2	2	Stratified-wavy	0.66	5.54% at 100% air extraction
0.2	3	Stratified-wavy	0.644	4% at 100% air extraction
0.2	4	Stratified-wavy	0.68	10% at 100% air extraction
0.2	5	Stratified-wavy	0.642	9.3% at 100% air extraction
0.2	6	Stratified-wavy	0.63	5% at 100% air extraction
0.2	7	Stratified-wavy	0.65	15% at 100% air extraction
0.2	8	Strati fied-wavy	0.63	14.5% at 100% air extraction
0.2	9	Stratified-wavy	0.60	11,2% at 100% air extraction
0.2	10	Semi-slug	0.55	16.1% at 86.5% air extraction
0.2	11	Semi-slug	0.34	12.2% at 88% air extraction
0.2	12	Slug	0.17	19.5% at 89% air extraction
0.2	13	Slug	0.15	16% at 90% air extraction
0.2	14	Semi-slug	0.50	20% at 87.3% air extraction
0.2	15	Semi-slug	0.24	17% at 89.9% air extraction
0.2	16	Slug	0.14	21% at 89% air extraction
0.2	17	Slug	0.13	17.5% at 89% air extraction
0.2	18	Semi-slug	0.58	16% at 87% air extraction
0.2	19	Slug	0.30	30% at 87.8% air extraction
0.2	20	Slug	0.16	24% at 87% air extraction

Table 5 Coded empirical correlations to predict the fraction of liquid *L'* in the side arm, obtained through regression analysis by Ejaz et al. [2]

References	Coded empirical correlations	Eqs. no
[3, 4]	$\sqrt{L'} = +0.7984 + 0.1081V_R + 0.0286D_R - 0.2302\theta + 0.321M_S$ $- 0.0388D_R M_{SR} - 0.1748V_R^2 + 0.224\theta^2 - 0.0788M_{SR}^2$	(6)
[5]	$L' = -0.7773 + 0.2943V_R - 0.0107D_R + 0.6479M_{SR} + 0.2215V_R M_{SR}$ $- 0.1297V_R^2$	(7)
[6]	$\sqrt{L'} = 12.38 - 0.0281V_R + 0.0656D_R + 18.64M_{SR} - 0.086V_R^2$ $+ 0.0585D_R^2 + 7.06M_{SR}^2$	(8)
[7]	$\sqrt{L'} = 0.5375 + 0.0916V_R + 0.1143D_R - 0.4598\theta + 0.2017M_{SR}$ $+ 0.0837V_R D_R - 0.0588V_R\theta + 0.0232D_R\theta + 0.1053\theta M_{SR}$	(9)
	$L'_{Peak} = 0.7429 + 0.3475D_R + 0.1483V_R - 0.3199\theta + 0.3407D_R V_R$	(10)
[8]	$\sqrt{L'} = 1.41 + 0.025V_R + 0.9084D_R + 1.11P + 0.9397D_R P - 0.0482V_R^2$ $-0.1028D_R^2 - 0.1743P^2$	(11)

Table 6 Uncoded or actual empirical correlations for replicability of the collected data [2]

References	Un-coded empirical correlations	Eqs. no
[3, 4]	$\sqrt{L'} = -0.042034 + 0..000105V_R + 0.269256D_R - 0.004264\theta + 1.19001M_S$ $- 0.310055D_R M_{SR} - 8.9e^{-9}V_R^2 + 0.000004\theta^2 - 0.315389M_{SR}^2$	(12)
[5]	$L' = -0.072131 + 0.000032V_R - 0.072666D_R$ $+ 0.968318M_{SR} + 0.00005V_R M_{SR} - 1.28e^{-9}V_R^2$	(13)
[6]	$\sqrt{L'} = 0.239372 - 0.0000353V_R + 0.497472D_R + 0.357676M_{SR}$ $- 4.827e^{-7}V_R^2 + 0.521235D_R^2 + 0.233251M_{SR}^2$	(14)
[7]	$\sqrt{L'} = 0.125765 - 0.000578V_R + 0.146056D_R - 0.0006644\theta + 0.403342M_{SR}$ $+ 0.000995V_R D_R - 1.55e^{-6}V_R\theta + 0.001287D_R\theta + 0.002339\theta M_{SR}$	(15)
	$L'_{Peak} = 0.578206 + 0.018865D_R - 0.002711V_R$ $- 0.003555\theta + 0.003802D_R V_R$	(16)
[8]	$\sqrt{L'} = -2.19391 + 0.049743V_R - 6.9672D_R + 3.609P + 7.13109D_R P$ $- 0.005549V_R^2 - 0.642322D_R^2 - 1.60608P^2$	(17)

Bibliography

1. Q. Zeeshan, *Experimental Investigation of Two-Phase Separation in T-Junction with Combined Diameter Ratio*, Universiti Teknologi Petronas (2019)
2. F. Ejaz, W. Pao, H.M. Ali, Numerical investigation and prediction of phase separation in diverging T-junction. Int. J. Numer. Methods Heat Fluid Flow. (2022). https://doi.org/10.1108/HFF-12-2021-0782
3. S. Marti, O. Shoham, A unified model for stratified-wavy two-phase flow splitting at a reduced T-junction with an inclined branch arm. Int. J. Multiph. Flow **23**, 725–748 (1997). https://doi.org/10.1016/S0301-9322(97)82477-9
4. O. Shoham, S. Arirachakaran, J.P. Brill, Two-phase flow splitting in a horizontal reduced pipe tee. Chem. Eng. Sci. **44**, 2388–2391 (1989). https://doi.org/10.1016/0009-2509(89)85174-7
5. L.C. Walters, *Two-Phase Pressure Drop and Phase Distribution at Horizontal T-Junction: The effect of branch diameter*. University of Manitoba (1994)
6. B.J. Azzopardi, Effect of side arm diameter on phase split at T-junctions, Proc. - SPE Annu. Tech. Conf. Exhib. 2 (1999). https://doi.org/10.2523/56707-ms
7. E. Wren, Geometric effects on phase split at a large diameter T-junction, Phd. Diss. (2001)
8. A. Saieed, *Experimental investigation on the effect of diameter ratio on two-phase separation in a T-junction* (Universiti Teknologi Petronas, Malaysia, 2018)

F. Ejaz et al., *Two Phase Separation in the T-Junction*,
https://doi.org/10.1007/978-981-99-9354-3

SPRINGER NATURE

GPSR Compliance

The European Union's (EU) General Product Safety Regulation (GPSR) is a set of rules that requires consumer products to be safe and our obligations to ensure this.

If you have any concerns about our products, you can contact us on ProductSafety@springernature.com

In case Publisher is established outside the EU, the EU authorized representative is:

Springer Nature Customer Service Center GmbH
Europaplatz 3
69115 Heidelberg, Germany

The manufacturer's authorised representative in the EU is Springer Nature Customer Service Centre GmbH, Europaplatz 3, 69115 Heidelberg, Germany. If you have any concerns regarding our products, please contact ProductSafety@springernature.com

Printed and bound by CPI Group (UK) Ltd, Croydon, CR0 4YY
01/10/2026
02225877-0002